T0187338

The Anxious Generation

JONATHAN HAIDT

The Anxious Generation

How the Great Rewiring of
Childhood Is Causing an
Epidemic of Mental Illness

ALLEN LANE
an imprint of
PENGUIN BOOKS

ALLEN LANE

UK | USA | Canada | Ireland | Australia
India | New Zealand | South Africa

Penguin Books is part of the Penguin Random House group of companies
whose addresses can be found at global.penguinrandomhouse.com.

First published in the United States of America by Penguin Press 2024
First published in Great Britain by Allen Lane 2024

012

Copyright © Jonathan Haidt, 2024

The moral right of the author has been asserted

Printed and bound in Great Britain by Clays Ltd, Elcograf S.p.A.

The authorized representative in the EEA is Penguin Random House Ireland,
Morrison Chambers, 32 Nassau Street, Dublin D02 YH68

A CIP catalogue record for this book is available from the British Library

HARDBACK ISBN: 978–0–241–64766–0
TRADE PAPERBACK ISBN: 978–0–241–69490–9

For the teachers and principals at P.S. 3,
LAB Middle School, Baruch Middle School,
and Brooklyn Technical High School, who have
devoted their lives to nurturing children,
including mine.

CONTENTS

Part 4
COLLECTIVE ACTION
FOR HEALTHIER CHILDHOOD

Introduction

GROWING UP ON MARS

Suppose that when your first child turned ten, a visionary billionaire whom you've never met chose her to join the first permanent human settlement on Mars. Her academic performance—plus an analysis of her genome, which you don't remember giving consent for—clinched her a spot. Unbeknownst to you, she had signed herself up for the mission because she loves outer space, and, besides, all of her friends have signed up. She begs you to let her go.

Before saying no, you agree to learn more. You learn that the reason they're recruiting children is that they adapt better to the unusual conditions of Mars than adults, particularly the low gravity. If children go through puberty and its associated growth spurt on Mars, their bodies will be permanently tailored to it, unlike settlers who come over as adults. At least that's the theory. It is unknown whether Mars-adapted children would be able to return to Earth.

You find other reasons for fear. First, there's the radiation. Earth's flora and fauna evolved under the protective shield of the magnetosphere, which blocks or diverts most of the solar wind, cosmic rays, and other streams of harmful particles that bombard our planet. Mars doesn't have such a shield, so a far greater number of ions would shoot through the DNA of each cell in your daughter's body. The project's planners have built protective shields for the Mars settlement based on studies of adult astronauts, who have a slightly elevated risk of cancer after spending a year in space.[1] But children are at an even higher risk, because their cells are developing and diversifying more rapidly and would experience higher rates of cellular damage. Did the planners take this into account? Did they do any research on child safety at all? As far as you can tell, no.

And then there's gravity. Evolution optimized the structure of every creature over eons for the gravitational force on our particular planet. From birth onward, each creature's bones, joints, muscles, and cardiovascular system develop in response to the unchanging one-way pull of gravity. Removing this constant pull profoundly affects our bodies. The muscles of adult astronauts who spend months in the weightlessness of space become weaker, and their bones become less dense. Their body fluids collect in places where they shouldn't, such as the brain cavity, which puts pressure on the eyeballs and changes their shape.[2] Mars has gravity, but it's only 38% of what a child would experience on Earth. Children raised in the low-gravity environment of Mars would be at high risk of developing deformities in their skeletons, hearts, eyes, and brains. Did the planners take this vulnerability of children into account? As far as you can tell, no.

So, would you let her go?

Of course not. You realize this is a completely insane idea—sending children to Mars, perhaps never to return to Earth. Why would any parent allow it? The company behind the project is racing to stake its claim to Mars before any rival company. Its leaders don't seem to know anything about child development and don't seem to care about children's safety. Worse still: *The company did not require proof of parental permission.* As long as a child checks a box stating she has obtained parental permission, she can blast off to Mars.

No company could ever take our children away and endanger them without our consent, or they would face massive liabilities. Right?

AT THE TURN OF THE MILLENNIUM, TECHNOLOGY COMPANIES BASED ON the West Coast of the United States created a set of world-changing products that took advantage of the rapidly growing internet. There was a widely shared sense of techno-optimism; these products made life easier, more fun, and more productive. Some of them helped people to connect and communicate, and therefore it seemed likely they would be a boon to the growing number of emerging democracies. Coming soon after the fall of the Iron Curtain, it felt like the dawn of a new age. The founders of these companies were hailed as heroes, geniuses, and global benefactors who, like Prometheus, brought gifts from the gods to humanity.

But the tech industry wasn't just transforming life for adults. It began transforming life for children too. Children and adolescents had been watching a lot of television since the 1950s, but the new technologies were far more portable, personalized, and engaging than anything that came before. Parents discovered this truth early, as I did in 2008, when my two-year-old son mastered the touch-and-swipe interface of my first iPhone. Many parents were relieved to find that a smartphone or tablet could keep a child happily engaged and quiet for hours. Was this safe? Nobody knew, but because everyone else was doing it, everyone just assumed that it must be okay.

Yet the companies had done little or no research on the mental health effects of their products on children and adolescents, and they shared no data with researchers studying the health effects. When faced with growing evidence that their products were harming young people, they mostly engaged in denial, obfuscation, and public relations campaigns.[3] Companies that strive to maximize "engagement" by using psychological tricks to keep young people clicking were the worst offenders. They hooked children during vulnerable developmental stages, while their brains were rapidly rewiring in response to incoming stimulation. This included social media companies, which inflicted their greatest damage on girls, and video game companies and pornography sites, which sank their hooks deepest into boys.[4] By designing a firehose of addictive content that entered through kids' eyes and ears, and by displacing physical play and in-person socializing, these companies have rewired childhood and changed human development on an almost unimaginable scale. The most intense period of this rewiring was 2010 to 2015, although the story I will tell begins with the rise of fearful and overprotective parenting in the 1980s and continues through the COVID pandemic to the present day.

What legal limits have we imposed on these tech companies so far? In the United States, which ended up setting the norms for most other countries, the main prohibition is the Children's Online Privacy Protection Act (COPPA), enacted in 1998. It requires children under 13 to get parental consent before they can sign a contract with a company (the terms of service) to give away their data and some of their rights when they open an account. That set the effective age of "internet adulthood" at 13, for reasons that had little to do with children's safety or mental health.[5] But the wording of the law doesn't require companies to verify ages; as long as a child checks a box to assert that she's old enough (or puts in the right fake birthday), she can go almost anywhere on the internet without her parents' knowledge or consent. In fact, 40% of American children under 13 have created Instagram accounts,[6] yet there has been no update of federal laws since 1998. (The U.K., on the other hand, has taken some initial steps, as have a few U.S. states.[7])

A few of these companies are behaving like the tobacco and vaping industries, which designed their products to be highly addictive and then skirted laws limiting marketing to minors. We can also compare them to the oil companies that fought against the banning of leaded gasoline. In the mid-20th century, evidence began to mount that the hundreds of thousands of tons of lead put into the atmosphere *each year*, just by drivers in the United States, were interfering with the brain development of tens of millions of children, impairing their cognitive development and increasing rates of antisocial behavior. Even still, the oil companies continued to produce, market, and sell it.[8]

Of course, there is an enormous difference between the big social media companies today and, say, the big tobacco companies of the mid-20th century: Social media companies are making products that are useful for adults, helping them to find information, jobs, friends, love, and sex; making shopping and political organizing more efficient; and making life easier in a thousand ways. Most of us would be happy to live in a world with no tobacco, but social media is far more valuable, helpful, and even beloved by many adults. Some adults have problems with addiction to social media and other online activities, but as with tobacco, alcohol, or gambling we generally leave it up to them to make their own decisions.

The same is not true for minors. While the reward-seeking parts of the brain mature earlier, the frontal cortex—essential for self-control, delay of gratification, and resistance to temptation—is not up to full capacity until the mid-20s, and preteens are at a particularly vulnerable point in development. As they begin puberty, they are often socially insecure, easily swayed by peer pressure, and easily lured by any activity that seems to offer social validation. We don't let preteens buy tobacco or alcohol, or enter casinos. The costs of using social media, in particular, are high for adolescents, compared with adults, while the benefits are minimal. Let children grow up on Earth first, before sending them to Mars.

THIS BOOK TELLS THE STORY OF WHAT HAPPENED TO THE GENERATION born after 1995,[9] popularly known as Gen Z, the generation that follows the millennials (born 1981 to 1995). Some marketers tell us that Gen Z ends with the birth year 2010 or so, and they offer the name Gen Alpha for the children born after that, but I don't think that Gen Z—the anxious generation—will have an end date until we change the conditions of childhood that are making young people so anxious.[10]

Thanks to the social psychologist Jean Twenge's groundbreaking work, we know that what causes generations to differ goes beyond the *events* children experience (such as wars and depressions) and includes *changes in the technologies* they used as children (radio, then television, personal computers, the internet, the iPhone).[11] The oldest members of Gen Z began puberty around 2009, when several tech trends converged: the rapid spread of high-speed broadband in the 2000s, the arrival of the iPhone in 2007, and the new age of hyper-viralized social media. The last of these was kicked off in 2009 by the arrival of the "like" and "retweet" (or "share") buttons, which transformed the social dynamics of the online world. Before 2009, social media was most useful as a way to keep up with your friends, and with fewer instant and reverberating feedback functions it generated much less of the toxicity we see today.[12]

A fourth trend began just a few years later, and it hit girls much harder than boys: the increased prevalence of posting images of oneself, after smartphones added front-facing cameras (2010) and Facebook acquired Instagram (2012), boosting its popularity. This greatly expanded the number of adolescents posting carefully curated photos and videos of their lives for their peers and strangers, not just to see, but to judge.

Gen Z became the first generation in history to go through puberty with a portal in their pockets that called them away from the people nearby and into an alternative universe that was exciting, addictive, unstable, and—as I will show—unsuitable for children and adolescents. Succeeding socially in that universe required them to devote a large part of their consciousness—perpetually—to managing what became their online brand. This was now necessary to gain acceptance from peers, which is the oxygen of adolescence, and to avoid online shaming, which is the nightmare of adolescence. Gen Z teens got sucked into spending many hours of each day scrolling through the shiny happy posts of friends, acquaintances, and distant influencers. They watched increasing quantities of user-generated videos and streamed entertainment, offered to them by autoplay and algorithms that were designed to keep them online as long as possible. They spent far less time playing with, talking to, touching, or even making eye contact with their friends and families, thereby reducing their participation in embodied social behaviors that are essential for successful human development.

The members of Gen Z are, therefore, the test subjects for a radical new way of growing up, far from the real-world interactions of small communities in which humans evolved. Call it the Great Rewiring of Childhood. It's as if they became the first generation to grow up on Mars.

THE GREAT REWIRING IS NOT JUST ABOUT CHANGES IN THE TECHNOLO-gies that shape children's days and minds. There's a second plotline here: the well-intentioned and disastrous shift toward overprotecting children and restricting their autonomy in the real world. Children need a great deal of free play to thrive. It's an imperative that's evident across all mammal species. The small-scale challenges and setbacks that happen during play are like an inoculation that prepares children to face much larger challenges later. But for a variety of historical and sociological reasons, free play began to decline in the 1980s, and the decline accelerated in the 1990s. Adults in the United States, the U.K., and Canada increasingly began to assume that if they ever let a child walk outside unsupervised, the child would attract kidnappers and sex offenders. Unsupervised outdoor play declined at the same time that the personal computer became more common and more inviting as a place for spending free time.*

I propose that we view the late 1980s as the beginning of the transition from a "play-based childhood" to a "phone-based childhood," a transition that was not complete until the mid-2010s, when most adolescents had their own smartphone. I use "phone-based" broadly to include all of the internet-connected personal electronics that came to fill young people's time, including laptop computers, tablets, internet-connected video game consoles, and, most important, smartphones with millions of apps.

---

* There is good evidence that the trends in overprotection, technology use, and mental health I describe happened in largely similar ways and at the same time in all of the countries of the Anglosphere: the United States, the U.K., Canada, Australia, and New Zealand (see Rausch & Haidt, 2023, March). I believe they are happening in most or all of the developed Western nations, although with variations based on the degree of individualism, social integration, and other cultural variables. I am collecting studies from other parts of the world and will be writing about trends in those countries on the *After Babel* Substack.

When I speak of a play-based or phone-based "childhood," I'm using that term broadly too. I mean it to include both children and adolescents (rather than having to write out "phone-based childhood and adolescence"). Developmental psychologists often mark the transition between childhood and adolescence as being the onset of puberty, but because puberty arrives at different times for different kids, and because it has been shifting younger in recent decades, it is no longer correct to equate adolescence to the teen years.[13] This is how age will be categorized in the rest of this book:

- **Children:** 0 through 12.
- **Adolescents:** 10 through 20.
- **Teens:** 13 through 19.
- **Minors:** Everyone who is under 18. I'll also use the word "kids" sometimes, because it sounds less formal and technical than "minors."

The overlap between children and adolescents is intentional: Kids who are 10 to 12 are between childhood and adolescence, and are often called tweens for that reason. (This period is also known as early adolescence.) They are as playful as younger children, yet they are beginning to develop the social and psychological complexities of adolescents.

As the transition from play-based to phone-based childhood proceeded, many children and adolescents were perfectly happy to stay indoors and play online, but in the process they lost exposure to the kinds of challenging physical and social experiences that all young mammals need to develop basic competencies, overcome innate childhood fears, and prepare to rely less on their parents. Virtual interactions with peers do not fully compensate for these experiential losses. Moreover, those whose playtime and social lives moved online found themselves increasingly wandering through adult spaces, consuming adult content, and interacting with adults in ways that are often harmful to minors. So even while parents worked to eliminate risk and freedom in the real world, they generally, and often unknowingly, granted full independence in the virtual world, in part because most found it difficult to understand what was going on there, let alone know what to restrict or how to restrict it.

My central claim in this book is that these two trends—*overprotection in the real world and underprotection in the virtual world*—are the major reasons why children born after 1995 became the anxious generation.

A FEW NOTES ABOUT TERMINOLOGY. WHEN I TALK ABOUT THE "REAL world," I am referring to relationships and social interactions characterized by four features that have been typical for millions of years:

1. They are *embodied*, meaning that we use our bodies to communicate, we are conscious of the bodies of others, and we respond to the bodies of others both consciously and unconsciously.
2. They are *synchronous*, which means they are happening at the same time, with subtle cues about timing and turn taking.
3. They involve primarily *one-to-one or one-to-several communication*, with only one interaction happening at a given moment.
4. They take place within communities that have a *high bar for entry and exit*, so people are strongly motivated to invest in relationships and repair rifts when they happen.

In contrast, when I talk about the "virtual world," I am referring to relationships and interactions characterized by four features that have been typical for just a few decades:

1. They are *disembodied*, meaning that no body is needed, just language. Partners could be (and already are) artificial intelligences (AIs).
2. They are heavily *asynchronous*, happening via text-based posts and comments. (A video call is different; it is synchronous.)
3. They involve a substantial number of *one-to-many communications*, broadcasting to a potentially vast audience. Multiple interactions can be happening in parallel.

4. They take place within communities that have a *low bar for entry and exit*, so people can block others or just quit when they are not pleased. Communities tend to be short-lived, and relationships are often disposable.

In practice, the lines blur. My family is very much real world, even though we use FaceTime, texting, and email to keep in touch. Conversely, a relationship between two scientists in the 18th century who knew each other only from an exchange of letters was closer to a virtual relationship. The key factor is the commitment required to make relationships work. When people are raised in a community that they cannot easily escape, they do what our ancestors have done for millions of years: They learn how to manage relationships, and how to manage themselves and their emotions in order to keep those precious relationships going. There are certainly many online communities that have found ways to create strong interpersonal commitments and a feeling of belonging, but in general, when children are raised in multiple mutating networks where they don't need to use their real names and they can quit with the click of a button, they are less likely to learn such skills.

THIS BOOK HAS FOUR PARTS. THEY EXPLAIN THE MENTAL HEALTH TRENDS among adolescents since 2010 (part 1); the nature of childhood and how we messed it up (part 2); the harms that result from the new phone-based childhood (part 3); and what we must do now to reverse the damage in our families, schools, and societies (part 4). Change is possible, if we can act together.

Part 1 has a single chapter laying out the facts about the decline in teen mental health and wellbeing in the 21st century, showing how devastating the rapid switch to a phone-based childhood has been. The decline in mental health is indicated by a sharp rise in rates of anxiety, depression, and self-harm, beginning in the early 2010s, which hit girls hardest. For boys, the story is more complicated. The increases are often smaller (except for suicide rates), and they sometimes begin a bit earlier.

Part 2 gives the backstory. The mental health crisis of the 2010s has its roots in the rising parental fearfulness and overprotection of the 1990s. I show how smartphones, along with overprotection, acted like "experience blockers," which made it difficult for children and adolescents to get the embodied social experiences they needed most, from risky play and cultural apprenticeships to rites of passage and romantic attachments.

In part 3, I present research showing that a phone-based childhood disrupts child development in many ways. I describe four foundational harms: sleep deprivation, social deprivation, attention fragmentation, and addiction. I then zoom in on girls* to show that social media use does not just *correlate* with mental illness; it *causes* it, and I lay out the empirical evidence showing multiple ways that it does so. I explain how boys came to their poor mental health by a slightly different route. I show how the Great Rewiring contributed to their rising rates of "failure to launch"—that is, to make the transition from adolescence to adulthood and its associated responsibilities. I close part 3 with reflections on how a phone-based life changes us all—children, adolescents, and adults—by bringing us "down" on what I can only describe as a spiritual dimension. I discuss six ancient spiritual practices that can help us all live better today.

In part 4, I lay out what we can and must do now. I offer advice, based on research, for what tech companies, governments, schools, and parents can do to break out of a variety of "collective action problems." These are traps long studied by social scientists in which an individual acting

---

* A note about gender: Girls and boys use different platforms (on average) in different ways, and they experience different patterns of mental health and mental illness, so a substantial part of this book (particularly chapters 6 and 7) looks at trends and processes separately for girls and boys. It is noteworthy that an increasing number of Gen Z youth identify as non-binary. Several studies indicate that the mental health of non-binary youth is even worse than for their male and female peers (see Price-Feeney et al., 2020). Research on this group remains scarce, both historically and currently. I hope that future studies will explore how these technologies distinctly affect non-binary youth. Most of the research I present in this book applies to all adolescents. For example, the four foundational harms affect adolescents regardless of their gender identity.

alone faces high costs, but if people can coordinate and act together, they can more easily choose actions that are better for all in the long run.

As a professor at New York University who teaches graduate and undergraduate courses, and who speaks at many high schools and colleges, I have found that Gen Z has several great strengths that will help them drive positive change. The first is that they are not in denial. They want to get stronger and healthier, and most are open to new ways of interacting. The second strength is that they want to bring about systemic change to create a more just and caring world, and they are adept at organizing to do so (yes, using social media). In the last year or so, I've been hearing about an increasing number of young people who are turning their attention to the ways the tech industry exploits them. As they organize and innovate, they'll find new solutions beyond those I propose in this book, and they'll make them happen.

I AM A SOCIAL PSYCHOLOGIST, NOT A CLINICAL PSYCHOLOGIST OR A MEDIA studies scholar. But the collapse of adolescent mental health is an urgent and complex topic that we can't understand from any one disciplinary perspective. I study morality, emotion, and culture. Along the way I've picked up some tools and perspectives that I'll bring to the study of child development and adolescent mental health.

I have been active in the field of positive psychology since its birth in the late 1990s, immersed in research on the causes of happiness. My first book, *The Happiness Hypothesis*, examines 10 "great truths" that ancient cultures East and West discovered about how to live a flourishing life.

Based on that book, I taught a course called Flourishing when I was a professor of psychology at the University of Virginia (until 2011), and I teach versions of it now at NYU's Stern School of Business, to undergraduates and to MBA students. I have seen the rising levels of anxiety and device addiction as my students have changed from millennials using flip phones to Gen Z using smartphones. I have learned from their candor in discussing their mental health challenges and their complex relationships with technology.

My second book, *The Righteous Mind*, lays out my own research on the evolved psychological foundations of morality. I explore the reasons why good people are divided by politics and religion, paying special attention to people's needs to be bound into moral communities that give them a sense of shared meaning and purpose. This work prepared me to see how online social networks, which can be useful for helping adults achieve their goals, may not be effective substitutes for real-world communities within which children have been rooted, shaped, and raised for hundreds of thousands of years.

But it was my third book that led me directly to the study of adolescent mental health. My friend Greg Lukianoff was among the first to notice that something had changed very suddenly on college campuses as students began engaging in exactly the distorted thinking patterns that Greg had learned to identify and reject when he learned CBT (cognitive behavioral therapy) after a severe episode of depression in 2007. Greg is a lawyer and the president of the Foundation for Individual Rights and Expression, which has long helped students defend their rights against censorious campus administrators. In 2014, he saw something strange happening: Students themselves began demanding that colleges protect them from books and speakers that made them feel "unsafe." Greg thought that universities were somehow *teaching* students to engage in cognitive distortions such as catastrophizing, black-and-white thinking, and emotional reasoning, and that this could actually be *causing* students to become depressed and anxious. In August 2015, we presented this idea in an *Atlantic* essay titled "The Coddling of the American Mind."

We were only partially correct: Some college courses and new academic trends[14] were indeed inadvertently teaching cognitive distortions. But by 2017 it had become clear that the rise of depression and anxiety was happening in many countries, to adolescents of all educational levels, social classes, and races. On average, people born in and after 1996 were different, psychologically, from those who had been born just a few years earlier.

We decided to expand our *Atlantic* article into a book with the same title. In it, we analyzed the causes of this mental illness crisis, drawing on Jean Twenge's 2017 book, *iGen*. At the time, however, nearly all evidence was correlational: Soon after teens got iPhones, they started getting more depressed. The heaviest users were also the most depressed, while those who spent more time in face-to-face activities, such as on sports teams and in religious communities, were the healthiest.[15] But given that correlation is not proof of causation, we cautioned parents not to take drastic action on the basis of existing research.

Now, as I write in 2023, there's a lot more research—experimental as well as correlational—showing that social media harms adolescents, especially girls going through puberty.[16] I have also discovered, while doing the research for this book, that the causes of the problem are broader than I had initially thought. It's not just about smartphones and social media; it's about a historic and unprecedented transformation of human childhood. The transformation is affecting boys as much as girls.

WE HAVE MORE THAN A CENTURY OF EXPERIENCE IN MAKING THE REAL world safe for kids. Automobiles became popular in the early 20th century and tens of thousands of children died in them until eventually we mandated seat belts (in the 1960s) and then car seats (in the 1980s).[17] When I was in high school in the late 1970s, many of my fellow students smoked cigarettes, which they could easily buy from vending machines. Eventually, America banned those machines, inconveniencing adult smokers, who then had to purchase cigarettes from a store clerk who could verify their age.[18]

Over the course of many decades, we found ways to protect children while mostly allowing adults to do what they want. Then quite suddenly, we created a virtual world where adults could indulge any momentary whim, but children were left nearly defenseless. As evidence mounts that phone-based childhood is making our children mentally unhealthy, socially isolated, and deeply unhappy, are we okay with that trade-off? Or will we eventually realize, as we did in the 20th century, that we sometimes need to protect children from harm even when it inconveniences adults?

I'll offer many ideas for reforms in part 4, all of which aim to reverse the two big mistakes we've made: overprotecting children in the real world (where they need to learn from vast amounts of direct experience) and underprotecting them online (where they are particularly vulnerable during puberty). The suggestions I offer are based on the research I present in parts 1 through 3. Since the research findings are complicated and some of them are contested among researchers, I will surely be wrong on some points, and I will do my best to correct any errors by updating the online supplement for the book. Nonetheless, there are four reforms that are so important, and in which I have such a high degree of confidence, that I'm going to call them foundational. They would provide a foundation for healthier childhood in the digital age. They are:

1. *No smartphones before high school.* Parents should delay children's entry into round-the-clock internet access by *giving only basic phones* (phones with limited apps and no internet browser) before ninth grade (roughly age 14).

2. *No social media before 16.* Let kids get through the most vulnerable period of brain development before connecting them to a firehose of social comparison and algorithmically chosen influencers.

3. *Phone-free schools.* In all schools from elementary through high school, students should store their phones, smartwatches, and any other personal devices that can send or receive texts in phone lockers or locked pouches during the school day. That is the only way to free up their attention for each other and for their teachers.

4. *Far more unsupervised play and childhood independence.* That's the way children naturally develop social skills, overcome anxiety, and become self-governing young adults.

These four reforms are not hard to implement—if many of us do them at the same time. They cost almost nothing. They will work even if we never get help from our legislators. If most of the parents and schools in a community were to enact all four, I believe they would see substantial improvements in adolescent mental health within two years. Given that

AI and spatial computing (such as Apple's new Vision Pro goggles) are about to make the virtual world far more immersive and addictive, I think we'd better start today.

WHILE WRITING *THE HAPPINESS HYPOTHESIS*, I CAME TO HAVE GREAT respect for ancient wisdom and the discoveries of previous generations. What would the sages advise us today about managing our phone-based lives? They'd tell us to get off our devices and regain control of our minds. Here is Epictetus, in the first century CE, lamenting the human tendency to let others control our emotions:

> If your body was turned over to just anyone, you would doubtless take exception. Why aren't you ashamed that you have made your mind vulnerable to anyone who happens to criticize you, so that it automatically becomes confused and upset?[19]

Anyone who checks their "mentions" on social media, or has ever been thrown for a loop by what somebody posted about them, will understand Epictetus's concern. Even those who are rarely mentioned or criticized, and who simply scroll through a bottomless feed featuring the doings, rantings, and goings-on of other people, will appreciate Marcus Aurelius's advice to himself, in the second century CE:

> Don't waste the rest of your time here worrying about other people— unless it affects the common good. It will keep you from doing anything useful. You'll be too preoccupied with what so-and-so is doing, and why, and what they're saying, and what they're think- ing, and what they're up to, and all the other things that throw you off and keep you from focusing on your own mind.[20]

Adults in Gen X and prior generations have not experienced much of a rise in clinical depression or anxiety disorders since 2010,[21] but many of us have become more frazzled, scattered, and exhausted by our new technologies and their incessant interruptions and distractions. As generative AI enables the production of super-realistic and fabricated photographs, videos, and news stories, life online is likely to get far more confusing.[22] It doesn't have to be that way; we can regain control of our own minds.

This book is not just for parents, teachers, and others who care for or about children. It is for anyone who wants to understand how the most rapid rewiring of human relationships and consciousness in human history has made it harder for all of us to think, focus, forget ourselves enough to care about others, and build close relationships.

The Anxious Generation is a book about how to reclaim human life for human beings in all generations.

Part 1

A TIDAL WAVE

Chapter 1

THE SURGE OF SUFFERING

When I talk with parents of adolescents, the conversation often turns to smartphones, social media, and video games. The stories parents tell me tend to fall into a few common patterns. One is the "constant conflict" story: Parents try to lay down rules and enforce limits, but there are just so many devices, so many arguments about why a rule needs to be relaxed, and so many ways around the rules, that family life has come to be dominated by disagreements about technology. Maintaining family rituals and basic human connections can feel like resisting an ever-rising tide, one that engulfs parents as well as children.

For most of the parents I talk to, their stories don't center on any diagnosed mental illness. Instead, there is an underlying worry that something unnatural is going on, and that their children are missing something—really, almost everything—as their online hours accumulate.

But sometimes the stories parents tell me are darker. Parents feel that they have lost their child. A mother I spoke with in Boston told me about the efforts she and her husband had made to keep their fourteen-year-old daughter, Emily,[1] away from Instagram. They could see the damaging effects it was having on her. To curb her access, they tried various programs to monitor and limit the apps on her phone. However, family life devolved into a constant struggle in which Emily eventually found ways around the restrictions. In one distressing episode, she got into her mother's phone, disabled the monitoring software, and threatened to kill herself if her parents reinstalled it. Her mother told me:

> It feels like the only way to remove social media and the smartphone from her life is to move to a deserted island. She attended summer camp for six weeks each summer where no phones were permitted—no electronics at all. Whenever we picked her up from camp she was her normal self. But as soon as she started using her phone again it was back to the same agitation and glumness. Last year I took her phone away for two months and gave her a flip phone and she returned to her normal self.

When I hear such stories about boys, they usually involve video games (and sometimes pornography) rather than social media, particularly when a boy makes the transition from being a casual gamer to a heavy gamer. I met a carpenter who told me about his 14-year-old son, James, who has mild autism. James had been making good progress in school before COVID arrived, and also in the martial art of judo. But once schools were shut down, when James was eleven, his parents bought him a PlayStation, because they had to find something for him to do at home.

At first it improved James's life—he really enjoyed the games and social connections. But as he started playing *Fortnite* for lengthening periods of time, his behavior began to change. "That's when all the depression, anger, and laziness came out. That's when he started snapping at us," the father told me. To address James's sudden change in behavior, he and his wife took all of his electronics away. When they did this, James showed withdrawal symptoms, including irritability and aggressiveness, and he refused to come out of his room. Although the intensity of his symptoms lessened after a few days, his parents still felt trapped: "We tried to limit his use, but he doesn't have any friends, other than those he communicates with online, so how much can we cut him off?"

No matter the pattern or severity of their story, what is common among parents is the feeling that they are trapped and powerless. Most parents don't want their children to have a phone-based childhood, but somehow the world has reconfigured itself so that any parent who resists is condemning their children to social isolation.

In the rest of this chapter, I'm going to show you evidence that something big is happening, something changed in the lives of young people in the early 2010s that made their mental health plunge. But before we immerse ourselves in the data, I wanted to share with you the voices of parents who feel that their children were in some sense swept away, and who are now struggling to get them back.

THE WAVE BEGINS

There was little sign of an impending mental illness crisis among adolescents in the 2000s.[2] Then, quite suddenly, in the early 2010s, things changed. Each case of mental illness has many causes; there is always a complex backstory involving genes, childhood experiences, and sociological factors. My focus is on why rates of mental illness went up in so many countries between 2010 and 2015 for Gen Z (and some late millennials) while older generations were much less affected. Why was there a synchronized international increase in rates of adolescent anxiety and depression?

Greg and I finished writing *The Coddling of the American Mind* in early 2018. Figure 1.1 is based on a graph that we included in our book, with data through 2016. I have updated it to show what has happened since. In a survey conducted every year by the U.S. government, teens are asked a series of questions about their drug use, along with a few questions about their mental health. Examples include asking whether you have experienced a long period of feeling "sad, empty, or depressed," or a long period in which you "lost interest and became bored with most of the things you usually enjoy"? Those who answered yes to more than five

Major Depression Among Teens

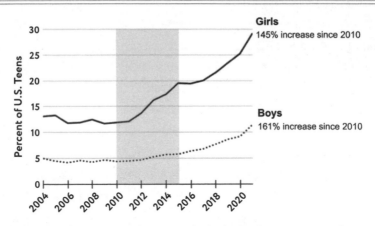

Figure 1.1. Percent of U.S. teens (ages 12–17) who had at least one major depressive episode in the past year, by self-report based on a symptom checklist. This was figure 7.1 in *The Coddling of the American Mind*, now updated with data beyond 2016. (Source: U.S. National Survey on Drug Use and Health.)[3]

out of nine questions about symptoms of major depression are classified as being highly likely to have suffered from a "major depressive episode" in the past year.

You can see a sudden and very large upturn in major depressive episodes, beginning around 2012. (In Figure 1.1, and in most of the graphs to follow, I have added a shaded area to make it easy for you to judge whether or not something changed between 2010 and 2015, which is the period I call "The Great Rewiring.") The increase for girls was much larger than the increase for boys in *absolute* terms (the number of additional cases since 2010), and a hockey stick shape jumps out more clearly. However, boys started at a lower level than girls, so in *relative* terms (the percent change since 2010, which I'll always use as the baseline), the increases were similar for both sexes—roughly 150%. In other words, depression became roughly *two and a half times more prevalent*. The increases happened across all races and social classes.[4] The data for 2020 was collected partly before and partly after COVID shutdowns, and by then one out of every four American teen girls had experienced a major depressive episode in the previous year. You can also see that things got worse in 2021;

the lines tilt more steeply upward after 2020. But the great majority of the rise was in place before the COVID pandemic.

THE NATURE OF THE SURGE

What on earth happened to teens in the early 2010s? We need to figure out *who* is suffering from *what*, beginning *when*. It is extremely important to answer these questions precisely, in order to identify the causes of the surge and to identify potential ways to reverse it. That is what my team set out to do, and this chapter will lay out in some detail how we came to our conclusions.

We found important clues to this mystery by digging into more data on adolescent mental health.[5] The first clue is that the rise is concentrated in disorders related to anxiety and depression, which are classed together in the psychiatric category known as *internalizing disorders*. These are disorders in which a person feels strong distress and experiences the symptoms *inwardly*. The person with an internalizing disorder feels emotions such as anxiety, fear, sadness, and hopelessness. They ruminate. They often withdraw from social engagement.

In contrast, externalizing disorders are those in which a person feels distress and turns the symptoms and responses *outward*, aimed at other people. These conditions include conduct disorder, difficulty with anger management, and tendencies toward violence and excessive risk-taking. Across ages, cultures, and countries, girls and women suffer higher rates of internalizing disorders, while boys and men suffer from higher rates of externalizing disorders.[6] That said, both sexes suffer from both, and both sexes have been experiencing more internalizing disorders and fewer externalizing disorders since the early 2010s.[7]

You can see the ballooning rates of internalizing disorders in figure 1.2, which shows the percentage of college students who said that they had received various diagnoses from a professional. The data comes from standardized surveys by universities, aggregated by the American College Health Association.[8] The lines for depression and anxiety start out much higher than all other diagnoses and then increase more than any

Mental Illness Among College Students

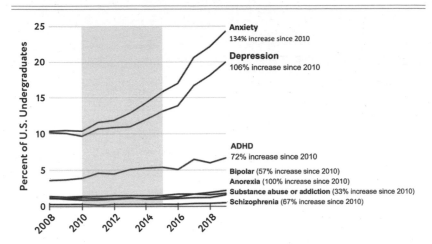

Figure 1.2. Percent of U.S. undergraduates with each of several mental illnesses. Rates of diagnosis of various mental illnesses increased in the 2010s among college students, especially for anxiety and depression. (Source: American College Health Association.)[9]

other in both relative and absolute terms. Nearly all of the increases in mental illness on college campuses in the 2010s came from increases in anxiety and/or depression.[10]

A second clue is that the surge is concentrated in Gen Z, with some spillover to younger millennials. You can see this in figure 1.3, which shows the percentage of respondents in four age-groups who reported feeling nervous in the past month "most of the time" or "all of the time." There is no trend for any of the four age-groups before 2012, but then the youngest group (which Gen Z begins to enter in 2014) starts to rise sharply. The next-older group (mostly millennials) rises too, though not as much, and the two oldest groups are relatively flat: a slight rise for Gen X (born 1965–1980) and a slight decrease for the baby boomers (born 1946–1964).

WHAT IS ANXIETY?

Anxiety is related to fear, but is not the same thing. The diagnostic manual of psychiatry (*DSM-5-TR*) defines fear as "the emotional response to

Anxiety Prevalence by Age

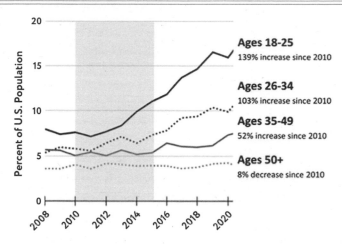

Figure 1.3. Percent of U.S. adults reporting high levels of anxiety by age group. (Source: U.S. National Survey on Drug Use and Health.)[11]

real or perceived imminent threat, whereas anxiety is anticipation of future threat."[12] Both can be healthy responses to reality, but when excessive, they can become disorders.

Anxiety and its associated disorders seem to be the defining mental illnesses of young people today. Across a variety of mental health diagnoses, you can see that anxiety rates rose most in figure 1.2, with depression following closely behind. A 2022 study of more than 37,000 high school students in Wisconsin found an increase in the prevalence of anxiety from 34% in 2012 to 44% in 2018, with larger increases among girls and LGBTQ teens.[13] A 2023 study of American college students found that 37% reported feeling anxious "always" or "most of the time," while an additional 31% felt this way "about half the time." This means that *only one-third of college students* said they feel anxiety less than half the time or never.[14]

Fear is arguably the most important emotion for survival across the animal kingdom. In a world rife with predators, those with lightning-fast responses were more likely to pass on their genes. In fact, quick responses to threats are so important that the brains of mammals can trigger a fear response before information from the eyes has even made it to the visual centers in the back of the brain for full processing.[15] This is why we can feel a wave of fear, or jump out of the way of an oncoming car, before we're even conscious of what we're looking at. Fear is an alarm bell connected to a rapid response system. Once the threat is over, the alarm stops ringing, stress hormones stop flowing, and the feeling of fear subsides.

While fear triggers the full response system at the moment of danger, anxiety triggers parts of the same system when a threat is merely perceived as possible. It is healthy to be anxious and on alert when one is in a situation where there really could be dangers lurking. But when our alarm bell is on a hair trigger so that it is frequently activated by ordinary events—including many that pose no real threat—it keeps us in a perpetual state of distress. This is when ordinary, healthy, temporary anxiety turns into an anxiety disorder.

It is also important to note that our alarm bell did not just evolve as a response to physical threats. Our evolutionary advantage came from our larger brains and our capacity to form strong social groups, thus making us particularly attuned to *social* threats such as being shunned or shamed. People—and particularly adolescents—are often more concerned about the threat of "social death" than physical death.

Anxiety affects the mind and body in multiple ways. For many, anxiety is felt in the body as tension or tightness and as discomfort in the abdomen and chest cavity.[16] Emotionally, anxiety is experienced as dread, worry, and, after a while, exhaustion. Cognitively, it often becomes difficult to think clearly, pulling people into states of unproductive rumination and provoking cognitive distortions that are the focus of cognitive behavioral therapy (CBT), such as catastrophizing, overgeneralizing, and black-and-white thinking. For those with anxiety disorders, these distorted thinking patterns often elicit uncomfortable physical symptoms, which then induce feelings of fear and worry, which then trigger more anxious thinking, perpetuating a vicious cycle.

The second most common psychological disorder among young people today is depression, as you can see in figure 1.2. The main psychiatric category here is called major depressive disorder (MDD). Its two key symptoms are depressed mood (feeling sad, empty, hopeless) and a loss of interest or pleasure in most or all activities.[17] "How weary, stale, flat and unprofitable, seem to me all the uses of this world," said Hamlet,[18] immediately after lamenting God's prohibition against "self-slaughter." For a diagnosis of MDD, these symptoms must be consistently present for at least two weeks. They are often accompanied by physical symptoms, including significant weight loss or weight gain, sleeping far less or far more than normal, and fatigue. They are also accompanied by disordered thinking, including an inability to concentrate, dwelling on one's transgressions or failings (causing feelings of guilt) and the many cognitive distortions that CBT tries to counteract. People experiencing a depressive disorder are likely to think about suicide because it feels like their current suffering will never end, and death is an end.

An important feature of depression for this book is its link to social relationships. People are more likely to become depressed when they become (or feel) more socially disconnected, and depression then makes people less interested and able to seek out social connection. As with anxiety, there is a vicious circle. So I'll be paying close attention to friendship and social relationships in this book. We'll see that a play-based childhood strengthens them, while a phone-based childhood weakens them.

I am not generally prone to anxiety or depression, yet I have suffered from prolonged anxiety, requiring medication, during three periods of my life. One included a diagnosis of major depression. So I can, to an extent, sympathize with what many young people are going through. I know that adolescents with anxiety or depressive disorders can't just "snap out of it" or decide to "toughen up." These disorders are caused by a combination of genes (some people are more predisposed to them), thought patterns (which can be learned and unlearned), and social or environmental conditions. But because genes didn't change between 2010 and 2015, we must figure out what thought patterns and social/environmental conditions changed to cause this tidal wave of anxiety and depression.

IT'S NOT REAL, IS IT?

Many mental health experts were initially skeptical that these large increases in anxiety and depression reflected real increases in mental illness. The day after we published *The Coddling of the American Mind*, an
essay appeared in *The New York Times* with the headline "The Big Myth
About Teenage Anxiety."[19] In it, a psychiatrist raised several important
objections to what he saw as a rising moral panic around teens and
smartphones. He pointed out that most of the studies showing a rise in
mental illness were based on "self-reports," like the data in figure 1.2. A
change in self-reports does not necessarily mean that there is a change
in underlying rates of mental illness. Perhaps young people just became
more willing to self-diagnose or more willing to talk honestly about their
symptoms? Or perhaps they started to mistake mild symptoms of anxiety for a mental disorder?

Was the psychiatrist right to be skeptical? He was certainly right
that we need to look at multiple indicators to know if mental illness really is increasing. A good way to do that is to look at changes in measures
not self-reported by teens. For example, many studies chart changes in
the number of adolescents brought in for emergency psychiatric care, or

Emergency Room Visits for Self-Harm

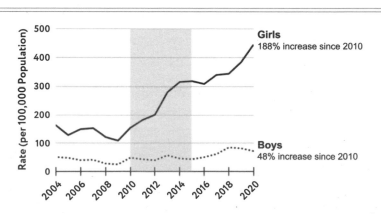

Figure 1.4. The rate per 100,000 in the U.S. population at which adolescents (ages
10–14) are treated in hospital emergency rooms for nonfatal self-injury. (Source: U.S.
Centers for Disease Control, National Center for Injury Prevention and Control.)[20]

admitted to hospitals each year because they deliberately harmed themselves. This can either be in a suicide attempt, commonly done by overdosing on medications, or in what is called nonsuicidal self-injury (NSSI), often done by cutting oneself without the intent to die. Figure 1.4 shows the data for visits to emergency rooms in the United States, and it shows a pattern similar to the rising rates of depression that we saw in figure 1.1, especially for girls.

The rate of self-harm for these young adolescent girls nearly *tripled* from 2010 to 2020. The rate for older girls (ages 15–19) doubled, while the rate for women over 24 actually went *down* during that time (see online supplement).[21] So whatever happened in the early 2010s, *it hit preteen and young teen girls harder than any other group*. This is a major clue. Acts of intentional self-harm in figure 1.4 include both nonfatal suicide attempts, which indicate very high levels of distress and hopelessness, and NSSI, such as cutting. The latter are better understood as coping behaviors that some people (especially girls and young women) use to manage debilitating anxiety and depression.

Adolescent suicide in the United States shows a time trend generally similar to depression, anxiety, and self-harm, although the period of rapid increase begins a few years earlier. Figure 1.5 shows the suicide

Suicide Rates for Younger Adolescents

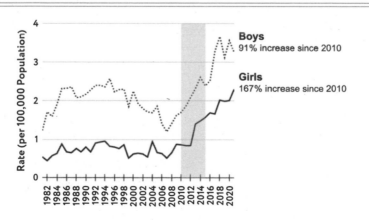

Figure 1.5. Suicide rates for U.S. adolescents, ages 10–14. (Source: U.S. Centers for Disease Control, National Center for Injury Prevention and Control.)[22]

rate, expressed as the number of children aged 10–14, per 100,000 such children in the U.S. population, who died by suicide in each year.[23] For suicide, the rates are nearly always higher for boys than for girls in Western nations, while attempted suicides and nonsuicidal self-harm are higher for girls, as we saw above.[24]

Figure 1.5 shows that the suicide rate for young adolescent girls began to rise in 2008, with a surge in 2012, after having bounced around within a limited range since the 1980s. From 2010 to 2021, the rate increased 167%. This too is a clue guiding us to ask: *What changed for preteen and younger teen girls in the early 2010s?*

The rapid increases in rates of self-harm and suicide, in conjunction with the self-report studies showing increases in anxiety and depression, offer a strong rebuttal to those who were skeptical about the existence of a mental health crisis. I am not saying that *none* of the increase in anxiety and depression is due to a greater willingness to report these conditions (which is a good thing) or due to some adolescents who began to pathologize normal anxiety and discomfort (which is not a good thing). But the pairing of self-reported suffering with behavioral changes tells us that something big changed in the lives of adolescents in the early 2010s, perhaps beginning in the late 2000s.

SMARTPHONES AND THE CREATION OF GEN Z

The arrival of the smartphone changed life for everyone after its introduction in 2007. Like radio and television before it, the smartphone swept the nation and the world. Figure 1.6 shows the percentage of American homes that had purchased various communication technologies over the last century. As you can see, these new technologies spread quickly, always including an early phase where the line seems to go nearly straight up. That's the decade or so in which "everyone" seems to be buying it.

Figure 1.6 shows us something important about the internet era: It came in two waves. The 1990s saw a rapid increase in the paired technologies of personal computers and internet access (via modem, back then), both of which could be found in most homes by 2001. Over the next

Communication Technology Adoption

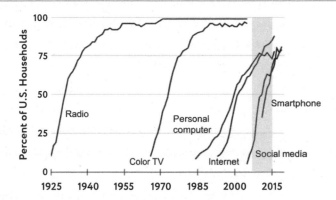

Figure 1.6. The share of U.S. households using specific technologies. The smartphone was adopted faster than any other communication technology in history. (Source: Our World in Data.)[25]

10 years, there was no decline in teen mental health.[26] Millennial teens, who grew up playing in that first wave, were slightly happier, on average, than Gen X had been when they were teens. The second wave was the rapid increase in the paired technologies of social media and the smartphone, which reached a majority of homes by 2012 or 2013. That is when girls' mental health began to collapse, and when boys' mental health changed in a more diffuse set of ways.

Of course, teens had cell phones since the late 1990s, but they were "basic" phones with no internet access, often known at the time as flip phones because the most popular design could be flipped open with a flick of the wrist. Basic phones were mostly useful for communicating directly with friends and family, one-on-one. You could call people, and you could text them using cumbersome thumb presses on a numeric keypad. Smartphones are very different. They connect you to the internet 24/7, they can run millions of apps, and they quickly became the home of social media platforms, which can ping you continually throughout the day, urging you to check out what everyone is saying and doing. This kind of connectivity offers few of the benefits of talking directly with friends. In fact, for many young people, it's poisonous.[27]

There are several sources for data on the early smartphone era. A

2012 report on cell phone ownership from Pew Research found that in 2011, 77% of American teens had a phone but *just 23% had a smartphone.*[28] That means most teens had to access social media using a computer. Often it was their parents' computer or the family computer, so they had limited privacy and access, and there was no easy way to get online when away from home. In the United States, laptop computers became increasingly common in this period, as did high-speed internet, so some teens started gaining increased access to the internet even before they got their own smartphones.

But it wasn't until teens got smartphones that they could be online *all the time*, even when away from home. According to a survey of U.S. parents conducted by the nonprofit Common Sense Media, by 2016, *79% of teens owned a smartphone*, as did 28% of children between the ages of 8 and 12.[29]

As adolescents got smartphones, they began spending more time in the virtual world. A 2015 Common Sense report found that teens with a social media account reported spending about two hours a day on social media, and teens overall reported spending an average of nearly seven hours a day of leisure time (not counting school and homework) on screen media, which includes playing video games and watching videos on Netflix, YouTube, or pornography sites.[30] A 2015 report by Pew Research[31] confirms these high numbers: One out of every four teens said that they were online "almost constantly." By 2022, that number had nearly doubled, to 46%.[32]

These "almost constantly" numbers are startling and may be the key to explaining the sudden collapse of adolescent mental health. These extraordinarily high rates suggest that even when members of Gen Z are not on their devices and *appear* to be doing something in the real world, such as sitting in class, eating a meal, or talking with you, a substantial portion of their attention is monitoring or worrying (being anxious) about events in the social metaverse. As the MIT professor Sherry Turkle wrote in 2015 about life with smartphones, "We are forever elsewhere."[33] This is a profound transformation of human consciousness and relationships, and it occurred, for American teens, between 2010 and 2015. This is the birth of the phone-based childhood. It marks the definitive end of the play-based childhood.

An important detail in this story is that the iPhone 4 was introduced in June 2010.[34] It was the first iPhone with a front-facing camera, which made it far easier to take photos and videos of oneself. Samsung offered one on its Galaxy S that same month. That same year, Instagram was created as an app that could be used only on smartphones. For the first few years, there was no way to use it on a desktop or laptop.[35] Instagram had a small user base until 2012, when it was purchased by Facebook. Its user base then grew rapidly (from 10 million near the end of 2011[36] to 90 million by early 2013[37]). We might therefore say that the smartphone and selfie-based social media ecosystem that we know today emerged in 2012, with Facebook's purchase of Instagram following the introduction of the front-facing camera. By 2012, many teen girls would have felt that "everyone" was getting a smartphone and an Instagram account, and everyone was comparing themselves with everyone else.

Over the next few years the social media ecosystem became even more enticing with the introduction of ever more powerful "filters" and editing software within Instagram and via external apps such as Facetune. Whether she used filters or not, the reflection each girl saw in the mirror got less and less attractive relative to the girls she saw on her phone.

While girls' social lives moved onto social media platforms, boys burrowed deeper into the virtual world as they engaged in a variety of digital activities, particularly immersive online multiplayer video games, YouTube, Reddit, and hardcore pornography—all of which became available anytime, anywhere, for free, right on their smartphones.

With so many new and exciting virtual activities, many adolescents (and adults) lost the ability to be fully present with the people around them, which changed social life for everyone, even for the small minority that did not use these platforms. That is why I refer to the period from 2010 to 2015 as the Great Rewiring of Childhood. Social patterns, role models, emotions, physical activity, and even sleep patterns were fundamentally recast, for adolescents, over the course of just five years. The daily life, consciousness, and social relationships of 13-year-olds with iPhones in 2013 (who were born in 2000) were profoundly different from those of 13-year-olds with flip phones in 2007 (who were born in 1994).

AREN'T THEY RIGHT TO BE ANXIOUS AND DEPRESSED?

When I present these findings in public, someone often objects by saying something like "Of course Gen Z is depressed; just look at the state of the world in the 21st century! It begins with the 9/11 attacks, the wars in Afghanistan and Iraq, and the global financial crisis. They're growing up with global warming, school shootings, political polarization, inequality, and ever-rising student loan debt. You point to 2012 as the pivotal year? That was the year of the Sandy Hook Elementary School shooting!"[38]

The 2021 book *Generation Disaster* offered exactly this argument for the mental health problems of Gen Z.[39] But while I agree that the 21st century is off to a bad start, the timing does not support the argument that Gen Z is anxious and depressed *because* of objective facts about rising national or global threats.

Even *if* we were to accept the premise that the events from 9/11 through the global financial crisis had substantial effects on adolescent mental health, they would have most heavily affected the millennial generation (born 1981 through 1995), who found their happy childhood world shattered and their prospects for upward mobility reduced. But this *did not* happen; their rates of mental illness did not worsen during their teenage years. Also, had the financial crisis and other economic concerns been major contributors, adolescent mental health in the United States would have plummeted in 2009, during the darkest year of the financial crisis, and it would have improved throughout the 2010s as the unemployment rate fell, the stock market rose, and the economy heated up. Neither of these trends is borne out in the data. In figure 1.7, I superimposed figure 1.1, about teen depression, on a graph of the U.S. unemployment rate, which spiked in 2008 and 2009 as companies threw employees overboard at the

Teenage Depression vs. Adult Unemployment

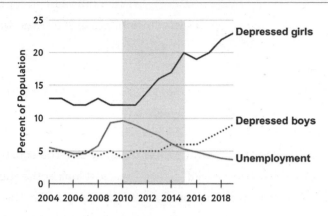

Figure 1.7. The U.S. unemployment rate (percent of adults in the labor market who are unemployed) fell continuously as the adolescent mental health crisis got worse. (Sources: U.S. Bureau of Labor Statistics and the U.S. National Survey Drug Use and Health.)[40]

start of the crisis. Unemployment then began a long, steady decline from 2010 to 2019, hitting a historic low of just 3.6% in early 2019.

There is just no way to pin the surge of adolescent anxiety and depression on any economic event or trend that I can find. Also, it's hard to see why an economic crisis would have harmed girls more than boys, and preteen girls more than everyone else.

The other explanation I often hear is that Gen Z is anxious and depressed because of climate change, which will affect their lives more than those of older generations. I do not deny that their concern is legitimate, but I want to point out that impending threats to a nation or generation (as opposed to an individual) do not historically cause rates of mental illness to rise. When countries are attacked, either by military force or by terrorism, citizens usually rally around the flag and each other. They are infused with a strong sense of purpose, suicide rates drop,[41] and researchers find that decades later, people who were teens during the start of the war show higher levels of trust and cooperation in lab experiments.[42] When young people rally together around a political cause, from opposing the Vietnam War in the 1960s through peak periods of earlier climate activism in the 1970s and 1990s, they become *ener-gized*, not dispirited or depressed. Every generation grows up during a disaster or under the threat of an impending disaster, from the Great Depression and World War II through threats of nuclear annihilation, environmental degradation, overpopulation, and ruinous national debt. People don't get depressed when they face threats collectively; they get depressed when they feel isolated, lonely, or useless. As I'll show in later chapters, this is what the Great Rewiring did to Gen Z.

Collective anxiety can bind people together and motivate them to take action, and collective action is thrilling, especially when it is carried out in person. Among previous generations, researchers often found that those engaged in political activism were happier and more energized than average. "There is something about activism itself that is beneficial for well-being," said Tim Kasser, coauthor of a 2009 study on college students, activism, and flourishing.[43] Yet more recent studies of young activists, including climate activists, find the opposite: Those who are politically active nowadays usually have *worse* mental health.[44] Threats and risks have always haunted the future, but the ways that young people are responding, with activism carried out mostly in the virtual world, seem to be affecting them very differently compared to previous generations, whose activism was carried out mostly in the real world.

The climate change hypothesis also fails to explain some of the demographic particularities here. Why do we usually see the biggest relative increases of anxiety and depression among preteen girls? Wouldn't an increased awareness of climate issues affect the oldest teens and college students more, because they are more aware of global and political events? It also fails to fit the timing: Why the spike in mental illness in the early 2010s, in so many countries? The Swedish climate activist Greta Thunberg (born 2003) galvanized young people around the world, but only after she addressed a UN Climate Change conference in 2018.

Everything may seem broken, but that was just as true when I was growing up in the 1970s and when my parents were growing up in the 1930s. It is the story of humanity. If world events played a role in the current mental health crisis, it's not because world events suddenly got worse around 2012; it's because world events were suddenly being pumped into adolescents' brains through their phones, not as news stories, but as social media posts in which other young people expressed their emotions about a collapsing world, emotions that are contagious on social media.

ALL OVER THE ANGLOSPHERE

One way to tell if American adolescents became anxious and depressed due to current events is to compare their mental health trends with those in other countries with different current events and different levels of cultural distance from the United States. Below I do this for a variety of countries: those that are culturally similar but had different major news events, such as Canada and the U.K.; those with different languages and cultures, namely the Nordic countries; and finally for 37 countries from around the world that participate in a survey of their 15-year-olds every three years. As I demonstrate, all of these show a similar pattern and timing: Something changed in the early 2010s.

Let's start with Canada, which shares much of its culture with the United States yet lacks many of America's potentially damaging sociological and economic features, such as high levels of economic insecurity. Canada has avoided America's frequent wars and high rates of violent crime. Canada also largely avoided the effects of the global financial crisis.[45] Yet even with all these advantages, adolescents in Canada experienced a sharp decline in mental health at the same time and in the same way as those in the United States.[46]

Figure 1.8 shows the percent of Canadian girls and women who reported that their mental health was either "excellent" or "very good." If you stopped collecting data in 2009, you'd conclude that the youngest group (aged 15–30) was the happiest, and you'd see no reason for concern. But in 2011 the line for the youngest women began to dip and then went into free fall while the line for the oldest group of women (aged 47

Excellent or Very Good Mental Health, Canadian Women

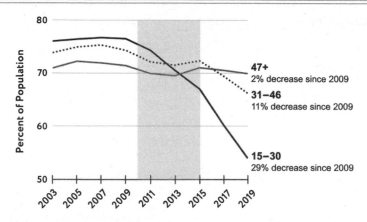

Figure 1.8. Percent of girls and women in Ontario, Canada, who reported that their mental health was either "excellent" or "very good." (Source: D. Garriguet [2021], Portrait of youth in Canada: Data report.)[47]

and up) didn't budge. The graph for boys and men shows the same pattern, though with a smaller decline. (You can find that graph and many more in the online supplement, which has a separate Google Doc for each chapter in this book. See anxiousgeneration.com/supplement.)

As in the United States, changes in behavior match changes in self-reported mental health. When we plot the rate of psychiatric emergency department visits for self-harm for Canadian teens, we find almost exactly the same pattern as for American teens in figure 1.4.[48]

It's the same story in the U.K., which has somewhat more cultural distance from the United States than does Canada. Nonetheless, its teens suffered in the same way and at the same time as those in the United States. Rates of anxiety and depression rose in the early 2010s, especially for girls.[49] And once again, we see the same sudden increase when we look at behavioral data. Figure 1.9 shows the rates at which U.K. teens deliberately harmed themselves, according to a study of medical records. As in the United States and Canada, something seems to have happened to British teens in the early 2010s that caused a sudden and large increase in the number of teens harming themselves.[50]

We see similar trends in the other major Anglosphere nations,

Self-Harm Episodes, U.K. Teens

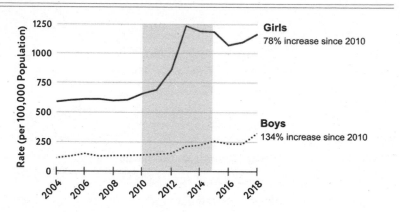

Figure 1.9. U.K. teens' (ages 13–16) self-harm episodes. (Source: Cybulski et. al., 2021, drawing from two databases of anonymized British medical records.)[51]

including Ireland, New Zealand, and Australia.[52] For example, figure 1.10 shows the rate at which Australian teens and young adults were admitted to hospitals for psychiatric emergencies. As in the other Anglo countries, if you stopped your data collection at the beginning of the Great Rewiring (2010), you'd see nothing, but by 2015 teens were in deep trouble.

Mental Health Hospitalizations, Australia

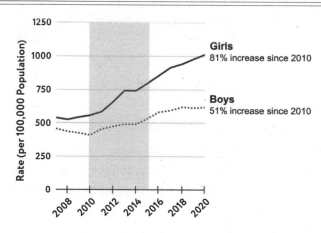

Figure 1.10. Rate at which Australian teens (ages 12–24) were kept in hospitals overnight for mental health reasons. (Source: Australia's Health 2022 Data Insights.)[53]

THE REST OF THE WORLD

In 2020, I hired Zach Rausch, a late millennial (born 1994) earning a master's degree in psychology, as a part-time research assistant. He quickly rose to become my full-time research partner for this book. Zach has gathered mental health data from all over the world and published several in-depth reports at the *After Babel* Substack (which we created to test out ideas for this book and my next one). In one report, Zach examined the five Nordic countries and found the same patterns as in the five Anglosphere countries. Figure 1.11 shows the percent of teens in Finland, Sweden, Denmark, Norway, and Iceland who reported high levels of psychological distress between 2002 and 2018.[54] The pattern is indistinguishable from those found repeatedly among the Anglo countries: If you cut off the graphs in 2010, at the start of the Great Rewiring, you see no sign of a problem. If you look at data through 2015, there's a big problem.

What about the world beyond the wealthy Anglosphere and Nordic nations? There are several global studies of *adult* mental health, but there are few global surveys of *adolescents*.[55] There is, however, a global educational survey called the Program for International Student Assessment,

High Psychological Distress, Nordic Nations

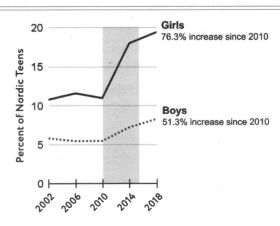

Figure 1.11. Percent of Nordic teens with high psychological distress (ages 11–15). (Source: Data from the Health Behavior in School Age Children Survey.)[56]

or PISA for short. Every three years since 2000, PISA surveys thousands of 15-year-olds and their parents in each of 37 participating countries. Nestled within hundreds of questions about their academic progress and their home life was a set of six questions about their feelings about school. These asked students to say how much they agreed with statements such as "I feel lonely at school," "I feel like an outsider (or left out of things) at school," and "I make friends easily at school" (which was reverse scored).[57]

Jean Twenge and I analyzed the responses to these six questions and plotted the aggregated scores since 2000 for all 37 countries.[58] Figure 1.12 shows those trends from four major world regions. After staying relatively flat from 2000 through 2012, reports of feeling lonely and friendless at school increased, in all regions except for Asia. Across the Western world, it seems that as soon as teens began carrying smartphones to school and using social media regularly, including during breaks between classes, they found it harder to connect with their fellow students. They were "forever elsewhere."

The 2008 global financial crisis did not cause this multinational

Alienation in School, Worldwide

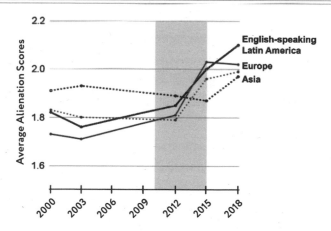

Figure 1.12. Worldwide school alienation scores over time (age 15). Note that the increase in school loneliness occurs in all regions other than Asia, mostly between 2012 and 2015. (These questions were not asked in the 2006 and 2009 surveys.) Scores range from 1 (low alienation) to 4 (high alienation). (Source: Twenge, Haidt et al. [2021]. Data from PISA.)[59]

increase in the 2010s, nor did American school shootings or American politics. The only plausible theory I have found that can explain the international decline in teen mental health is the sudden and massive change in the technology that teens were using to connect with each other.[60]

CHILDREN BORN IN THE LATE 1990S WERE THE FIRST GENERATION IN history who went through puberty in the virtual world. It's as though we sent Gen Z to grow up on Mars when we gave them smartphones in the early 2010s, in the largest uncontrolled experiment humanity has ever performed on its own children.

IN SUM

- Between 2010 and 2015, the social lives of American teens moved largely onto smartphones with continuous access to social media, online video games, and other internet-based activities. This Great Rewiring of Childhood, I argue, is the single largest reason for the tidal wave of adolescent mental illness that began in the early 2010s.

- The first generation of Americans who went through puberty with smartphones (and the entire internet) in their hands became more anxious, depressed, self-harming, and suicidal. We now call that generation Gen Z, in contrast to the millennial generation, which had largely finished puberty when the Great Rewiring began in 2010.

- The tidal wave of anxiety, depression, and self-harm hit girls harder than boys, and it hit preteen girls hardest of all.

- The mental health crisis has also hit boys. Their rates of depression and anxiety have also increased a lot, although usually not by as much as for girls. Boys' technology use and mental health difficulties are somewhat different from those of girls, as I'll show in chapter 7.

- Suicide rates in the United States began rising around 2008 for adolescent boys and girls; they rose much higher in the 2010s.

- The increase in suffering was not limited to the United States. The same pattern is seen at roughly the same time among teens in the U.K., Canada, and other major Anglosphere countries, and also in the five Nordic nations. Feelings of alienation in school rose after 2012 across the Western world. Data is less abundant in non-Western nations, and the patterns there are less clear.[61]

- No other theory has been able to explain why rates of anxiety and depression surged among adolescents in so many countries at the same time and in the same way. Other factors, of course, contribute to poor mental health, but the unprecedented rise between 2010 and 2015 cannot be explained by the global financial crisis, nor by any set of events that happened in the United States or in any other particular country.

How, exactly, does a phone-based childhood interfere with child development and produce or exacerbate mental illness? To answer that question, we must first consider what childhood is and what children need to do in order to develop into healthy adults. That is my goal in part 2. I'll tell the backstory to the Great Rewiring, which is the gradual loss—beginning in the 1980s—of the play-based childhood.

Part 2

———

THE BACKSTORY

THE DECLINE OF THE PLAY-BASED CHILDHOOD

Chapter 2

WHAT CHILDREN NEED TO DO IN CHILDHOOD

Imagine that you fell into a deep sleep on June 28, 2007—the day before the iPhone was released. Like Rip Van Winkle, the protagonist in an 1819 story by Washington Irving, you wake up 10 years later and look around. The physical world looks largely the same to you, but people are behaving strangely. Nearly all of them are clutching a small glass and metal rectangle, and anytime they stop moving, they assume a hunched position and stare at it. They do this the moment they sit down on a train, or enter an elevator, or stand in line. There is an eerie quiet in public places—even babies are silent, mesmerized by these rectangles. When you do hear people talking, they usually seem to be talking to themselves while wearing white earplugs.

I borrowed this thought experiment from my collaborator Tobias Rose-Stockwell and his wonderful book, *Outrage Machine*. Tobias uses this scenario to convey the transformation of the *adult* world. But the thought experiment applies even more powerfully to the world of late childhood and adolescence. In 2007, teens and many preteens were busy tapping out short texts on their phones, but texting in those days was cumbersome (press the 7 key four times to make an *s*). Most of their texts were with one person at a time, and most used their basic phones to arrange ways to meet up in person. Nobody wanted to spend three consecutive hours texting. After the Great Rewiring, however, it became common for adolescents to spend most of their waking hours interacting with a smartphone, consuming content from strangers as well as friends, playing mobile games, watching videos, and posting on social media. By 2015, adolescents had a lot less time and motivation to get together in person.[1]

What happens to child and adolescent development when daily life—especially social life—gets radically rewired in this way? Might the new phone-based childhood alter the complex interplay of biological, psychological, and cultural development? Might it block kids from doing some of the things they need to do in order to turn into healthy, happy, competent, and successful adults? To answer these questions, we need to step back and look at five important features of human childhood.

SLOW-GROWTH CHILDHOOD

Here's a strange fact about human beings: Our kids grow fast, then slow, then fast. If you plot human growth curves against those of chimpanzees, you see that chimps grow at a steady pace until they reach sexual maturity, at which point they reproduce.[2] And why not? If evolution is all about maximizing surviving offspring, wouldn't it be most adaptive to get to the reproduction part as fast as possible?

But human children wait. They grow rapidly for the first two years, slow down for the next seven to 10, and then undergo a rapid growth spurt during puberty before coming to a halt a few years later. Intriguingly, a child's brain is already 90% of its full size by around age 5.[3] When *Homo sapiens* emerged, its children were big-brained small-bodied weaklings who ran around the forest practically begging predators to eat them. Why did we evolve to have this long and risky childhood?

The primary reason is that we evolved into cultural creatures between 1 million and 3 million years ago, roughly when our genus—*Homo*—emerged from earlier hominid species. Culture, which includes tool making, profoundly reshaped our evolutionary path. To give just one example: As we began using fire to cook our food, our jaws and guts reduced in size because cooked foods are so much easier to chew and digest. Our brains grew larger because the race for survival was won no longer by the fastest or strongest but by those most adept at learning. Our planet-changing trait was the ability to *learn from each other* and tap into the common pool of knowledge our ancestors and community had stored. Chimpanzees do very little of this.[4] Human childhood extended to give children time to learn.

The evolutionary race to *learn the most* made it maladaptive to reach puberty as fast as possible. Rather, there was a benefit to slowing things down. The brain doesn't grow much in size during late childhood, but it is busy making new connections and losing old ones. As children seek out experiences and practice a range of skills, the neurons and synapses that are used infrequently fade away, while frequent connections solidify and quicken. In other words, evolution has provided humans an extended childhood that allows for a long period of learning the accumulated knowledge of one's society—a kind of cultural apprenticeship, during adolescence, before one is seen and treated as an adult.

But evolution didn't just lengthen childhood to make learning *possible*. It also installed three strong motivations to do things that make learning *easy and likely*: motivations for free play, attunement, and social learning. In the days of play-based childhood, the norm was that when school let out, children were out playing with each other, unsupervised, in ways that let them satisfy these motives. But in the transition to phone-based childhood, the designers of smartphones, video game systems, social media, and other addictive technologies lured kids into the virtual world, where they no longer got the full benefit of acting on these three motivations.

FREE PLAY

Play is the work of childhood,[5] and all young mammals have the same job: Wire up your brain by playing vigorously and often. Hundreds of studies on young rats, monkeys, and humans show that young mammals *want* to play, *need* to play, and come out socially, cognitively, and emotionally *impaired* when they are deprived of play.[6]

In play, young mammals learn the skills they will need to be successful as adults, and they learn in the way that neurons like best: from repeated activity with feedback from success and failure in a low-stakes environment. Kittens pounce clumsily on a piece of yarn that triggers specialized circuits in their visual cortex that evolved to make them very interested in anything that looks like a mouse's tail. Gradually, after many playful pounces, they'll become skilled mouse killers. Human toddlers clumsily run around and climb up, over, or into anything they can, until they become skilled at moving around a complex natural environment. With those basic skills mastered, they move on to more advanced multiplayer predator-prey games, such as tag, hide-and-seek, and sharks and minnows. As they get older still, verbal play—as in gossip, teasing, and joking around—gives them an advanced course in nuance, nonverbal cues, and instantaneous relationship repair when something they said fails to produce the desired response. Over time, they develop the social skills necessary for life in a democratic society, including self-governance, joint decision making, and accepting the outcome when you lose a contest. Peter Gray, a developmental psychologist at Boston College and a leading play researcher, says that "play requires suppression of the drive to dominate and enables the formation of long-lasting cooperative bonds."[7]

Gray defines "free play" as "activity that is freely chosen and directed by the participants and undertaken for its own sake, not consciously pursued to achieve ends that are distinct from the activity itself."[8] Physical play, outdoors and with other children of mixed ages, is the healthiest, most natural, most beneficial sort of play. Play with some degree of physical risk is essential because it teaches children how to look after themselves and each other.[9] Children can only learn how to *not* get hurt in situations where it is possible to get hurt, such as wrestling with a friend, having a pretend sword fight, or negotiating with another child to enjoy a seesaw when a failed negotiation can lead to pain in one's posterior, as well as embarrassment. When parents, teachers, and coaches get involved, it becomes less free, less playful, and less beneficial. Adults usually can't stop themselves from directing and protecting.

A key feature of free play is that *mistakes are generally not very costly*. Everyone is clumsy at first, and everyone makes mistakes every day. Gradually, from trial and error, and with direct feedback from playmates, elementary school students become ready to take on the greater social complexity of middle school. It's not homework that gets them ready, nor is it classes on handling their emotions. Such adult-led lessons may provide useful information, but information doesn't do much to shape a developing brain. Play does. This relates to a key CBT insight: Experience, not information, is the key to emotional development. It is in unsupervised, child-led play where children best learn to tolerate bruises, handle their emotions, read other children's emotions, take turns, resolve conflicts, and play fair. Children are intrinsically motivated to acquire these skills because they want to be included in the playgroup and keep the fun going.

This is why I have chosen the term "play-based childhood" as a central term in this book, to be contrasted with a "phone-based childhood." A play-based childhood is one in which kids spend the majority of their free time playing with friends in the real world as I defined it in the introduction: embodied, synchronous, one-to-one or one-to-several, and in groups or communities where there is some cost to join or leave so people invest in relationships. This is how childhood was among hunter-gatherers, according to anthropological reports gathered by Gray,[10] which means that human childhood evolved during a long period in which brain development "expected" an enormous amount of free play. Of course, many children have had (and some still have) a work-based childhood. Work-based childhood was widespread during the Industrial Revolution, which is why, eventually, the 1959 United Nations Declaration of the Rights of the Child named play as a basic human right: "The child shall have full opportunity for play and recreation, which should be directed to the same purposes as education."[11]

So you can see the problem when some adolescents start spending the majority of their waking hours on their phones (and other screens), sitting alone watching YouTube videos on auto-play or scrolling through bottomless feeds on Instagram, TikTok, and other apps. These interactions generally have the contrasting features of the virtual world: disembodied, asynchronous, one-to-many, and done either alone or in virtual groups that are easy to join and easy to leave.

Even if the content on these sites could somehow be filtered effectively to remove obviously harmful material, the addictive design of these platforms reduces the time available for face-to-face play in the real world. The reduction is so severe that we might refer to smartphones and tablets in the hands of children as *experience blockers*. Of course, a smartphone opens up worlds of *new* possible experiences, including video games (which are forms of play) and virtual long-distance friendships. But this happens at the cost of reducing the kinds of experiences humans evolved for and that they must have in abundance to become socially functional adults. It's as if we gave our infants iPads loaded with movies about walking, but the movies were so engrossing that kids never put in the time or effort to practice walking.

The way young people use social media is generally not much like free play. In fact, posting and commenting on social media sites is the *opposite* of Gray's definition. Life on the platforms forces young people to become their own brand managers, always thinking ahead about the social consequences of each photo, video, comment, and emoji they choose. Each action is not necessarily done "for its own sake." Rather, every public action is, to some degree, strategic. It is, in Peter Gray's phrase, "consciously pursued to achieve ends that are distinct from the activity itself." Even for kids who never post anything, spending time on social media sites can still be harmful because of the chronic social comparison, the unachievable beauty standards, and the enormous amount of time taken away from everything else in life.

Surveys show that unstructured time with friends plummeted in the exact years that adolescents moved from basic phones to smartphones—the early 2010s. Figure 2.1 shows the percentage of U.S. students (combining 8th, 10th, and 12th graders) who said that they meet up with their friends "almost every day."

For boys and for girls there was a slow decline in the 1990s and early

Meet Up with Friends Daily

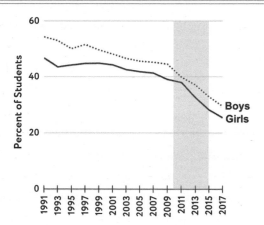

Figure 2.1. Percentage of U.S. students (8th, 10th, and 12th grade) who say that they meet up with their friends "almost every day" outside school.[12] (Source: Monitoring the Future. I explain how I use this important dataset in the endnotes.)[13]

2000s, which I'll discuss in the next chapter, followed by a faster decline in the 2010s. These accelerating declines are not just *evidence* of the Great Rewiring of Childhood, they *are* the Great Rewiring. Figure 2.1 shows us a generation moving away from the real world and into the virtual, thanks to the combination of smartphones, social media, multiplayer video games, and high-speed wireless internet.

ATTUNEMENT

Human children are wired to connect, in part by tuning and synchronizing their movements and emotions with others. Even before they can control their arms and legs, they engage adults in games of turn taking and shared emotion. Children respond with the most heart-opening peals of laughter when adults—who are themselves built to respond to cuteness with caretaking[14]—do whatever they can to make the baby laugh. This creates a mutually reinforcing feedback loop. Infants in the first weeks of life have enough muscular control to mimic a few facial expressions, and the many rounds of mutual gazing and face making are important means of fostering attachment between parents and children.[15]

Smartphones can disrupt this essential face-to-face interaction. Pew Research has found that 17% of American parents report they are *often* distracted by their phone when spending time with their child, with another 52% saying they are *sometimes* distracted.[16] Although new technologies have long distracted parents from their children, smartphones are uniquely effective at interfering with the bond between parent and child. With notifications constantly pinging and interrupting, some parents attend to their smartphones more than to their children, even when they are playing together.

When toddlers begin to speak, vast new possibilities for attunement open up. The social connections with parents and other caretakers grow deeper. Turn taking and good timing are essential social skills, and they begin to develop in these simple interactions: How long should I wait before I make the next funny face or give the next rhyme in the rhyming game we're developing? Each partner learns to read the other's facial expressions and emotions to get the timing right. Developmental psychologists refer to these sorts of interactions as *serve-and-return*, conveying the idea that social interactions are often like a game of tennis or ping-pong: You take turns, it's fun, there's unpredictability, and timing is essential.

Attunement practice is as essential for social development as movement and exercise are for physical development. According to the National Institute for Play,

Attunement forms the foundations for later emotional self-regulation. Children who are deprived of this joyful, mutually trusting social experience often face emotional difficulties and exhibit erratic behavior in their later years. They can have difficulty forming healthy attachments as adolescents and as adults they may be less able to cope with unexpected challenges, regulate emotions, make sound decisions when risk is involved, or learn to deal effectively as they enter into more and more complex social interactions.[17]

As children get older, they go beyond turn taking to find joy in perfect synchrony, doing the same thing at the same time as their partner. Girls especially come to delight in singing songs together, jumping rope together, or playing rhyming and clapping games (such as pat-a-cake) in which high-speed hand motions are perfectly matched between the partners while high-speed nonsense songs are sung at the same time. Such games have no explicit goal or way to win. They are pleasurable because they use the ancient power of synchrony to create communion between unrelated people.

Anthropologists have long noted that collective rituals are universally human. The European explorers of the 16th and 17th centuries found that on every continent, communities performed rituals in which everyone moved together to drumming, chanting, or beat-heavy music.[18] Such rituals were widely said to renew trust and mend frayed social relations. The great sociologist Émile Durkheim wrote about the "social electricity" generated by such rituals;[19] he thought rituals were essential for fostering a sense of communion and belonging.

Many experiments have now shown that synchronous movement has exactly these effects. In one study, small groups of college students were given headphones to wear and were asked to hold up a beer mug and sway along with the music that they heard. Half of the groups swayed in perfect harmony (because they were listening to the same music at the same time). Half were out of sync (because the music was delivered to their headphones that way). All groups then played a trust game in which a group makes the most money if they all cooperate across many rounds, but any one of them could earn more money by making the selfish choice on any single round. Groups that had moved in sync with each other trusted each other more, cooperated more, and made more money than those that had moved out of sync.[20]

Synchronous, face-to-face, physical interactions and rituals are a deep, ancient, and underappreciated part of human evolution. Adults enjoy them, and children need them for healthy development. Yet the major social media platforms draw children into endless hours of *asynchronous* interaction, which can become more like work than play. Most teens have accounts on multiple platforms, and those who use social media regularly spend *two hours a day* or more just on social media sites.[21] By 2014, nearly a third of teen girls were spending *over 20 hours a week* on social media sites. That's half of a full-time job—creating content for the platform and consuming content created by others. That is time no longer available for interacting with friends in person. The work is often joyless, yet many feel compelled to do it, lest they "miss out" on something or be excluded.[22] Eventually, for many, it becomes a mindless habit, something they turn to dozens of times each day. Such social labor creates shallow connections because it is asynchronous and public, unlike a face-to-face conversation, or a private phone call or video call. And the interactions are disembodied; they use almost no muscles, other than in the swiping and typing fingers. We are physical, embodied creatures who evolved to use our hands, facial expressions, and head movements as communication channels, responding in real time to the similar movements of our partners. Gen Z is learning to pick emojis instead.

The loss of attunement is a second way that social media alters the course of childhood (while also fraying the social fabric). Given the vast amounts of time now invested in asynchronous interaction rather than getting together with friends, is it any wonder that so many teens found themselves lonely and starved for connection starting in the early 2010s?

SOCIAL LEARNING

Once our ancestors became cultural creatures, a new evolutionary pressure arose that rewarded the best learners. That doesn't mean those who learn best in school from books and lectures. It means those children who best activated their innate desire to learn by *copying* and who then *picked the right people* to copy.

You might think that choosing role models is simple: Children should just copy their parents, right? But that turns out to be a losing strategy. There is no reason for a child to assume that her *own* parents happen to be the most skilled adults in the community, so why not search more widely? Also, children need to learn how to be a successful *older child* in their particular community, so children are particularly attentive to such models.

According to Rob Boyd and Pete Richerson, two of the leading scholars of gene-culture coevolution,[23] there are several "strategies" that won out over thousands of generations and became part of our evolved propensity for culture. The two that are most relevant for our discussion of social media are *conformist bias* and *prestige bias.*

The value of conformity is obvious: Doing whatever most people are doing is the safest strategy across a wide range of environments. It's particularly valuable when you are a newcomer to an existing society: When in Rome, do as the Romans do. So when a child starts at a new school, she is particularly likely to do whatever it is that most children seem to be doing. We sometimes call this peer pressure, but it can be quite strong even when nobody is exerting pressure of any kind. It may be more accurate to call it conformity attraction. When American children move from elementary school to middle school (around age 11), they often discover, as my kids did, that most of their classmates have an Instagram account, which makes them want one too. And once on Instagram, they quickly learn how most of the people they follow use the platform, which makes them prone to using it that way too.

In a real-life social setting, it takes a while—often weeks—to get a good sense for what the most common behaviors are, because you need to observe multiple groups in multiple settings. But on a social media platform, a child can scroll through a thousand data points in one hour (at three seconds per post), each one accompanied by numerical evidence (likes) and comments that show whether the post was a success or a failure.

Social media platforms are therefore *the most efficient conformity engines ever invented.* They can shape an adolescent's mental models of acceptable behavior in a matter of hours, whereas parents can struggle unsuccessfully for years to get their children to sit up straight or stop whining. Parents don't get to use the power of conformity bias, so they are often no match for the socializing power of social media.[24]

But there's an important learning strategy that goes beyond copying the majority: Detect prestige and then copy the prestigious. The major work on prestige bias was done by the evolutionary anthropologist Joe Henrich,[25] who was a student of Rob Boyd's. Henrich noted that the social hierarchies of nonhuman primates are based on dominance—the ability, ultimately, to inflict violence on others. But humans have an alternative ranking system based on *prestige*, which is willingly conferred by people to those they see as having achieved excellence in a valued domain of activity, such as hunting or storytelling back in ancient times.

People can perceive excellence for themselves, but it's more efficient to rely on the judgments of others. If most people say that Frank is the best archer in your community, and if you value archery, you'll "look up" to Frank even if you've never seen him shoot an arrow. Henrich argues that the reason people become so deferential (starstruck) toward prestigious people is that they are motivated to get close to prestigious people in order to maximize their own learning and raise their own prestige by association. Prestigious people, in turn, will allow some supplicants to get close to them because having a retinue (a group of devoted attendants and followers) is a reliable signal to the community of their high standing in the prestige rankings.

Platform designers in Silicon Valley directly targeted this psychological system when they quantified and displayed the success of every post (likes, shares, retweets, comments) and every user, whose followers are literally called followers. Sean Parker, one of the early leaders of Facebook, admitted in a 2017 interview that the goal of Facebook's and Instagram's founders was to create "a social-validation feedback loop . . . exactly the kind of thing that a hacker like myself would come up with, because you're exploiting a vulnerability in human psychology."[26] But when the programmers quantified prestige based on the clicks of others, they hacked our psychology in ways that have been disastrous for young people's social development. On social media platforms, the ancient link between excellence and prestige can be severed more easily than ever, so in following influencers who became famous for what they do in the virtual world, young people are often learning ways of talking, behaving, and emoting that may backfire in an office, family, or other real-world setting.

The rise of mass media in the 20th century initiated this decoupling of excellence and prestige. The phrase "famous for being famous" first became popular in the 1960s when it became possible for an ordinary person to rise in the public's consciousness not for having done anything important but simply for having been seen by millions on TV and then being talked about over a few news cycles.[27] The phrase was later applied to the socialite and model Paris Hilton in the early 2000s, although her fame still depended on coverage by the mainstream and tabloid press. It was one of Paris Hilton's closet organizers—Kim Kardashian—who redefined the phrase for the social media age. Kardashian pioneered a new path to high prestige that began with a sex tape that went public on the internet, which led to a reality TV show (*Keeping Up with the Kardashians*) that introduced her entire family to the public. In 2023, Kim had 364 million followers on Instagram, and her sister Kylie had 400 million.

Prestige-based social media platforms have hacked one of the most important learning mechanisms for adolescents, diverting their time, attention, and copying behavior away from a variety of role models with whom they could develop a mentoring relationship that would help them succeed in their real-world communities. Instead, beginning in the early 2010s, millions of Gen Z girls collectively aimed their most powerful learning systems at a small number of young women whose main excellence seems to be amassing followers to influence. At the same time, many Gen Z boys aimed their social learning systems at popular male influencers who offered them visions of masculinity that were also quite extreme and potentially inapplicable to their daily lives.

EXPECTANT BRAINS AND SENSITIVE PERIODS

Children express their desires to play, to attune with others, and to learn socially in different ways throughout the long cultural apprenticeship of their slow-growth childhood and their fast-growth puberty. Healthy brain development depends on getting the right experiences at the right age and in the right order.

In fact, brain development in mammals and birds is sometimes called "experience-expectant development"[28] because specific parts of the brain show increased malleability during periods of life when the animal is likely to have a specific kind of experience. The clearest example is the existence of "critical periods," which are windows of time in which a young animal *must* learn something, or it will be hard if not impossible to learn later. Ducks, geese, and many other water- or ground-dwelling birds have an evolved learning mechanism called imprinting that tells the babies which adult they must follow. They will follow whatever mother-sized object moves in their field of vision a set number of hours after hatching. Many psychology textbooks show the photo in Figure 2.2 of the ethologist Konrad Lorenz being trailed by a line of goslings, who

Figure 2.2. Baby geese who had imprinted on Konrad Lorenz's boots.[29]

had imprinted on his boots because he had walked around the goslings during their critical period. Later research showed that it is possible for the young geese to learn a new attachment after the window has closed, yet even then the first thing they imprinted on retains a strong pull.[30] It has been stamped into their brains forever.

Humans have few true "critical periods" with hard time limits, but we do seem to have several "sensitive periods," which are defined as periods in which it is very easy to learn something or acquire a skill, and outside of which it is more difficult.[31] Language learning is the clearest case. Children can learn multiple languages easily, but this ability drops off sharply during the first few years of puberty.[32] When a family moves to a new country, the kids who are 12 or younger will quickly become native speakers with no accent, while those who are 14 or older will probably be asked, for the rest of their lives, "Where are you from?"

There seems to be a similar sensitive period for cultural learning, which closes just a few years later—still during puberty. The Japanese anthropologist Yasuko Minoura studied the children of Japanese businessmen who had been transferred by their companies to live for a few years in California during the 1970s.[33] She wanted to know at what age America shaped their sense of self, their feelings, and their ways of interacting with friends, even after they returned to Japan. The answer, she found, was between ages 9 and 14 or 15. Those children who spent a few years in California during that sensitive period came to "feel American." If they returned to Japan at 15 or later, they had a harder time readjusting, or coming to "feel Japanese." Those who didn't arrive in America until age 15 had no such problems, because they never came to feel American, and those who returned to Japan well before 14 were able to readjust, because they were still in their sensitive period and could relearn Japanese ways. Minoura noted that "during the sensitive period, a cultural meaning system for interpersonal relationships appears to become a salient part of self-identity to which they are emotionally attached."[34]

So what happens to American children who generally get their first smartphone around the age of 11 and then get socialized into the cultures of Instagram, TikTok, video games, and online life for the rest of their teen years? The sequential introduction of age-appropriate experiences, tuned to sensitive periods and shared with same-age peers, had been the norm during the era of play-based childhood. But in a phone-based childhood, children are plunged into a whirlpool of adult content and experiences that arrive in no particular order. Identity, selfhood, emotions, and relationships will all be different if they develop online rather than in real life. What gets rewarded or punished, how deep friendships become, and above all what is *desirable*—all of these will be determined by the thousands of posts, comments, and ratings that the child sees each week. Any child who spends her sensitive period as a heavy user of social media will be shaped by the cultures of those sites. This may explain why Gen Z's mental health outcomes are so much worse than those of the millennials: Gen Z was the first generation to go through puberty and the sensitive period for cultural learning on smartphones.

This hypothesis about puberty is not just my own speculation; a recent British study found direct evidence that puberty is indeed a sensitive period for harm from social media. A team led by the psychologist Amy Orben analyzed two large British data sets and found that the negative correlation between social media use and satisfaction with life was larger for those in the 10–15 age group than for those in the 16–21 age group, or any other age bracket.[35] They also examined a large longitudinal study to see if British teens who increased their social media use in one year would report worse mental health in the *following* year's survey. For those in the peak years of puberty, which comes a bit earlier for girls, the answer was yes. For girls, the worst years for using social media were 11 to 13; for boys, it was 14 to 15.[36]

These results offer clear evidence that 13, which is the current (and unenforced) minimum age for opening an account on social media platforms, is too low. Thirteen-year-olds should not be scrolling through endless posts from influencers and other strangers when their brains are in such an open state, searching for exemplars to lock onto. They should be playing, synchronizing, and hanging out with their friends in person while leaving some room in the input streams to their eyes and ears for social learning from their parents, teachers, and other role models in their communities.

PUTTING THIS ALL TOGETHER, WE CAN NOW UNDERSTAND THOSE SHARP "elbows" in so many of the graphs in the previous chapter. Gen Z is the first generation to have gone through puberty hunched over smartphones and tablets, having fewer face-to-face conversations and shoulder-to-shoulder adventures with their friends. As childhood was rewired—especially between 2010 and 2015—adolescents became more anxious, depressed, and fragile. In this new phone-based childhood, free play, attunement, and local models for social learning are replaced by screen time, asynchronous interaction, and influencers chosen by algorithms. Children are, in a sense, deprived of childhood.

IN SUM

- Human childhood is very different from that of any other animal. Children's brains grow to 90% of full size by age 5, but then take a long time to configure themselves. This slow-growth childhood is an adaptation for cultural learning. Childhood is an apprenticeship for learning the skills needed for success in one's culture.

- Free play is as essential for developing social skills, like conflict resolution, as it is for developing physical skills. But play-based childhoods were replaced by phone-based childhoods as children and adolescents moved their social lives and free time onto internet-connected devices.

- Children learn through play to connect, synchronize, and take turns. They enjoy attunement and need enormous quantities of it. Attunement and synchrony bond pairs, groups, and whole communities. Social media, in contrast, is mostly asynchronous and performative. It inhibits attunement and leaves heavy users starving for social connection.

- Children are born with two innate learning programs that help them to acquire their local culture. Conformist bias motivates them to copy whatever seems to be most common. Prestige bias motivates them to copy whoever seems to be the most accomplished and prestigious. Social media platforms, which are engineered for engagement, hijack social learning and drown out the culture of one's family and local community while locking children's eyes onto influencers of questionable value.

- Social learning occurs throughout childhood, but there may be a sensitive period for cultural learning that spans roughly ages 9 to 15. Lessons learned and identities formed in these years are likely to imprint, or stick, more than at other ages. These are the crucial sensitive years of puberty. Unfortunately, they are also the years in which most adolescents in developed countries get their own phones and move their social lives online.

Chapter 3

DISCOVER MODE AND
THE NEED FOR RISKY PLAY

In recent decades, America and many other Western nations made two contradictory choices about children's safety, and both were wrong. We decided that the real world was so full of dangers that children should not be allowed to explore it without adult supervision, even though the risks to children from crime, violence, drunk drivers, and most other sources have dropped steeply since the 1990s.[1] At the same time, it seemed like too much of a bother to design and require age-appropriate guardrails for kids online, so we left children free to wander through the Wild West of the virtual world, where threats to children abounded.

To take one example of our shortsightedness, a powerful fear for many parents is that their child will fall into the hands of a sexual predator. But sex criminals nowadays spend most of their time in the virtual world because the internet makes it so much easier to communicate with children and to find and circulate sexual and violent videos involving children. To quote a 2019 *New York Times* article, "Tech companies are reporting a boom in online photos and videos of children being sexually abused—a record 45 million illegal images were flagged last year alone—exposing a system at a breaking point and unable to keep up with the perpetrators."[2] More recently, in 2023, *The Wall Street Journal* ran an exposé that showed how "Instagram connects pedophiles and guides them to content sellers via recommendation systems that excel at linking those who share niche interests."[3]

To offer another example: Isabel Hogben, a 14-year-old girl in Rhode Island, wrote an essay in *The Free Press* that demonstrated how American parents are focusing on the wrong threats:

> I was ten years old when I watched porn for the first time. I found myself on Pornhub, which I stumbled across by accident and returned to out of curiosity. The website has no age verification, no ID requirement, not even a prompt asking me if I was over 18. The site is easy to find, impossible to avoid, and has become a frequent rite of passage for kids my age. Where was my mother? In the next room, making sure I was eating nine differently colored fruits and vegetables on the daily. She was attentive, nearly a helicopter parent, but I found online porn anyway. So did my friends.

Hogben's essay is a succinct illustration of the principle that *we are overprotecting our children in the real world while underprotecting them online.* If we really want to keep our children safe, we should delay their entry into the virtual world and send them out to play in the real world instead.

Unsupervised outdoor play teaches children how to handle risks and challenges of many kinds. By building physical, psychological, and social competence, it gives kids confidence that they can face new situations, which is an inoculation against anxiety. In this chapter, I'll show that a healthy human childhood with a lot of autonomy and unsupervised play in the real world sets children's brains to operate mostly in "discover mode," with a well-developed attachment system and an ability to handle the risks of daily life. Conversely, when there is society-wide pressure on parents to adopt modern overprotective parenting, it sets children's brains to operate mostly in "defend mode," with less secure attachment and reduced ability to evaluate or handle risk. Let me explain what these terms mean, and why discover mode is one of the keys to helping the anxious generation.

DISCOVER MODE VERSUS DEFEND MODE

The environments that shaped hominid evolution over the last few million years were extraordinarily variable, with periods of safety and abundance alternating with periods of scarcity, danger, drought, and starvation.[4] Our ancestors needed psychological adaptations to help them thrive in both settings. The variability of our environments shaped and refined older brain networks into two systems that are specialized for those two kinds of situations. The behavioral *activation* system (or BAS) turns on when you detect opportunities, such as suddenly coming across a tree full of ripe cherries when you and your group are hungry.[5] You're flooded with positive emotions and shared excitement, your mouths may begin to water, and everyone is ready to go! I'll give BAS a more intuitive name: *discover mode.*[6]

The behavioral *inhibition* system (BIS), in contrast, turns on when threats are detected, such as hearing a leopard roar nearby as you're picking those cherries. You all stop what you are doing. Appetite is suppressed as your bodies flood with stress hormones and your thinking turns entirely to identifying the threat and finding ways to escape it. I'll refer to BIS as *defend mode.* For people with chronic anxiety, defend mode is chronically activated.

The two systems together form a mechanism for quickly adapting to changing conditions, like a thermostat that can activate either a heating system or a cooling system as the temperature fluctuates. Across species, the default setting of the overall system depends on the animal's evolutionary history and expected environment. Animals that evolved with little daily risk of sudden death (such as top predators in a food chain, or herbivores on an island with no predators) often seem serene and confident. They are willing to get close to humans. Their default setting is discover mode, although they will shift into defend mode if attacked. In contrast, animals such as rabbits and deer, which evolved in the presence of constant predation, are skittish; they are quick to bolt and run. Their default setting is defend mode, and they shift into discover mode only slowly and tentatively when they perceive that the environment is unusually safe.

In humans (and other highly sociable mammals, such as dogs), the default setting is a major contributor to their individual personality. People (and dogs) who go through life in discover mode (except when directly threatened) are happier, more sociable, and more eager for new experiences. Conversely, people (and dogs) who are chronically in defend mode are more defensive and anxious, and they have only rare moments of perceived safety. They tend to see new situations, people, and ideas as potential threats, rather than as opportunities. Such chronic wariness was adaptive in some ancient environments, and may still be today for children raised in unstable and violent settings. But being stuck in defend mode is an obstacle to learning and growth in the physically safe environments that surround most children in developed nations today.

STUDENTS ON THE DEFENSIVE

Discover mode fosters learning and growth. If we want to help young people thrive—at home, in school, and in the workplace—shifting them into discover mode may be the most effective change we can make. Let me lay out the differences between the modes as we might see them in a college student. Figure 3.1 shows what a student arriving at a university would look like if her childhood (and her genes) gave her a brain whose default setting was discover mode versus defend mode. It's obvious that students in discover mode will profit and grow rapidly from the bountiful intellectual and social opportunities of a university. Students who spend most of their time in defend mode will learn less and grow less.

This contrast explains the sudden change that happened on many college campuses around 2014. Figure 3.2 shows how the distribution of mental challenges changed as the first members of Gen Z arrived and

Two Basic Mindsets

Discover mode (BAS)	Defend mode (BIS)
• Scan for opportunities	• Scan for dangers
• Kid in a candy shop	• Scarcity mindset
• Think for yourself	• Cling to your team
• *Let me grow!*	• *Keep me safe!*

Figure 3.1. Discover mode versus defend mode, for a student arriving at a university.

the last members of the millennial generation began to graduate. The only disorders that rose rapidly were psychological disorders. Those disorders were overwhelmingly anxiety and depression.

As soon as Gen Z arrived on campus, college counseling centers were overwhelmed.[7] The previously exuberant culture of millennial students in discover mode gave way to a more anxious culture of Gen Z students in defend mode. Books, words, speakers, and ideas that caused little or no controversy in 2010 were, by 2015, said to be harmful, dangerous, or traumatizing. America's residential universities are not perfect, but they are among the safest, most welcoming and inclusive environ-

Self-Reported Disabilities, College Freshmen

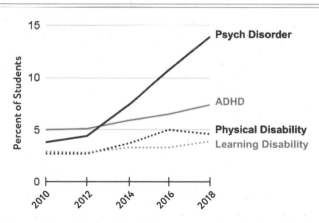

Figure 3.2. Percentage of U.S. college freshmen reporting various kinds of disabilities and disorders. (Source: Annual Freshman Survey, by UCLA's Higher Education Research Institute.)[8]

ments ever created for young adults. Yet campus culture changed around 2015, not just in the United States but also at British[9] and Canadian[10] universities. How could such a big change happen so quickly and internationally?

In the rest of this chapter, I'll show how the play-based childhood is nature's way of wiring up brains that tend toward discover mode, and how the phone-based childhood shifted a generation of children toward defend mode.

KIDS ARE ANTIFRAGILE

In the late 1980s, a grand experiment was launched in the Arizona desert. Biosphere 2 was (and still is) the largest attempt to build a closed artificial ecosystem, as a prelude to (someday) building self-sustaining ecosystems in outer space. Biosphere 2 was designed to support eight people, who would attempt to live within it for several years. All of the oxygen they breathed, the water they drank, and the food they ate was to be generated within the facility.

That goal was never reached. The complexity of biological interactions among species and social interactions among humans proved to be too much, but a great deal was learned from the multiple failures. For instance, many of the trees they planted to create a rain-forest ecosystem grew rapidly but then fell over before reaching maturity. The designers had not realized that young trees need wind to grow properly. When the wind blows, it bends the tree, which tugs at the roots on the windward side and compresses the wood on the other side. In response, the root system expands to provide a firmer anchor where it is needed, and the compressed wood cells change their structure to become stronger and firmer.

This altered cell structure is called reaction wood, or sometimes stress wood. Trees that are exposed to strong winds early in life become trees that can withstand even stronger winds when full grown. Conversely, trees that are raised in a protected greenhouse sometimes fall over from their own weight before they reach maturity.

Stress wood is a perfect metaphor for children, who also need to experience frequent stressors in order to become strong adults. The Biosphere trees illustrate the concept of "antifragility," a term coined by my NYU colleague Nassim Taleb in his 2012 book, *Antifragile: Things That Gain from Disorder.* Taleb noted that some things, like wineglasses, are fragile. We protect fragile things from shocks and threats because we know they cannot withstand even a gentle challenge, such as being knocked over on a dinner table. Other things are resilient, such as a plastic cup, which can withstand being knocked off the table. But resilient objects don't get better from getting dropped; they merely don't get worse.

Taleb coined the word "antifragile" to describe things that actually *need* to get knocked over now and then in order to become strong. I used the word "things," but there are very few inanimate objects that are antifragile. Rather, antifragility is a common property of complex systems that were designed (by evolution, and sometimes by people) to function in a world that is unpredictable.[11] The ultimate antifragile system is the immune system, which *requires* early exposure to dirt, parasites, and bacteria in order to set itself up in childhood. Parents who try to raise their children in a bubble of perfect hygiene are harming their children by blocking the development of their antifragile immune systems.

It's the same dynamic for what has been called the psychological immune system[12]—the ability of a child to handle, process, and get past frustrations, minor accidents, teasing, exclusion, perceived injustices, and normal conflicts without falling prey to hours or days of inner turmoil. There is no way to live with other humans without conflicts and deprivations. As the Stoics and Buddhists taught long ago, happiness cannot be reached by eliminating all "triggers" from life; rather, happiness comes from learning to deprive external events of the power to trigger negative emotions in you. In fact, the best parenting book[13] that my wife and I read when our children were toddlers urged us to look for opportunities to frustrate our children every day by laying out and enforcing the contingencies of life: *If you want to watch* Teletubbies, *you must first put away your toys. If you persist in doing that, you'll get a time-out. Yes, your sister got something you didn't, and that happens sometimes.*

Well-intentioned parents who try to raise their children in a bubble of satisfaction, protected from frustration, consequences, and negative emotions, may be harming their children. They may be blocking the development of competence, self-control, frustration tolerance, and emotional self-management. Several studies find that such "coddling" or "helicopter parenting" is correlated with later anxiety disorders, low self-efficacy (which is the inner confidence that one can do what is needed to reach one's goals), and difficulty adjusting to college.[14]

Children are intrinsically antifragile, which is why overprotected children are more likely to become adolescents who are stuck in defend mode. In defend mode, they're likely to learn less, have fewer close friends, be more anxious, and experience more pain from ordinary conversations and conflicts.

ANTIFRAGILE KIDS NEED RISKY PLAY TO STAY IN DISCOVER MODE

Antifragility is the key to solving many puzzles about human development, such as this one: Why do children add risk to their play? Why is it that once a skill is mastered, such as skateboarding down a gentle slope, a child will move on to a steeper slope, then a staircase, then perhaps the staircase railing? Why would children choose activities that pretty much guarantee that they'll get hurt, multiple times? Play researchers have long known the answer. As the Norwegian researchers Ellen Sandseter and Leif Kennair wrote in 2010, thrilling experiences have anti-phobic effects.[15]

Sandseter and Kennair begin with a puzzling fact long known in clinical psychology: Phobias are concentrated around a few animals and situations that kill almost nobody, such as snakes (even tiny ones), tightly enclosed places, the dark, public speaking, and heights. Conversely, very few people develop phobias to things that kill many modern people, including cars, opioids, knives, guns, and junk food. Furthermore, phobias in adults can rarely be traced to a bad experience in childhood.[16] In fact, kids who fall out of trees often turn into the adults who are least afraid of climbing trees.

We can resolve the puzzle by taking an evolutionary view. Common phobias evolved over millions of years of hunter-gatherer life, with some (such as snakes) being shared by other primates. We have an "evolved preparedness" to pay attention to some things, such as snakes, and to acquire a fear very easily from a single bad experience or from seeing others in our group show fear toward snakes. Conversely, as a child gains exposure, experience, and mastery, fear usually recedes.

As children become more competent, they become increasingly more *intrigued* by some of the things that had frightened them. They may approach them, look to adults and older kids for guidance, learn to distinguish the dangerous situations from the less dangerous ones, and eventually master their fears. As they do so, their fear turns into thrill and triumph. You can see the transition on a young child's face as he reaches out to touch a worm under a rock you just lifted up for him on a nature walk. You can see the mix of fear and fascination turning into a shriek of delight and disgust as he pulls his finger away, laughing. He did it! Now he'll be less afraid the next time he encounters a worm.

While I was writing this chapter, in the fall of 2022, my family got a puppy. Wilma is a small dog, and she weighed only seven pounds when we first started taking her for walks on the crowded sidewalks of New York City. At first she was visibly afraid of everything, including the parade of larger dogs, and she had trouble relaxing enough to "do her business."

Over time, she habituated somewhat, and I began to let her run off-leash, early mornings, in parks with other dogs. There too she was afraid at first, but the way she handled it made it seem as though she had read Sandseter and Kennair. She would approach much larger dogs, slowly, and then bolt away like lightning when they'd take a step toward her. Sometimes she'd run toward me for safety, but then her anti-phobic programming would kick in. Without slowing down, she'd execute a high-speed turn around my legs and sprint back toward the larger dog for

Figure 3.3. Wilma, age 7 months, executing a hairpin turn as her sprint toward a German shepherd sharply angled into a sprint away, which was followed by play position and more sprinting toward the larger dog. You can see the video of this interaction in the online supplement.

another round of thrills. She was experimenting to find the balance of joy and fear that she was ready for at that moment. By repeatedly cycling through discover and defend mode, she learned how to size up the intentions of other dogs and she developed her own abilities to engage in rough and joyful play, even as she occasionally got knocked over in a scramble of paws and tails.

Kids and puppies are thrill seekers. They are hungry for thrills, and they must get them if they are to overcome their childhood fears and wire up their brains so that discover mode becomes the default. Children need to swing and then jump off the swing. They need to explore forests and junkyards in search of novelty and adventure. They need to shriek with their friends while watching a horror movie or riding a roller coaster. In the process they develop a broad set of competences, including the ability to judge risk for themselves, take appropriate action when faced with risks, and learn that when things go wrong, even if they get hurt, they can usually handle it without calling in an adult.

Sandseter and Kennair define risky play as "thrilling and exciting forms of play that involve a risk of physical injury." (In a 2023 paper, expanding on their original work, they add that risky play also requires elements of uncertainty.[17]) They note that such play usually takes place outdoors, during free-play time rather than during activities organized by adults. Children choose to do activities that often lead to relatively harmless injuries, particularly bruises and cuts.

Sandseter and Kennair analyzed the kinds of risks that children seek out when adults give them some freedom, and they found six: heights (such as climbing trees or playground structures), high speed (such as swinging, or going down fast slides), dangerous tools (such as hammers and drills), dangerous elements (such as experimenting with fire), rough-and-tumble play (such as wrestling), and disappearing (hiding, wandering away, potentially getting lost or separated). These are the major types of thrills that children need. They'll get them for themselves unless adults stop them—which we did in the 1990s. Note that video games offer *none* of these risks, even though games such as *Fortnite* show avatars doing *all*

Figure 3.4. An overly dangerous playground in Dallas, Texas, year unknown.[18]

of them.[19] We are embodied creatures; children should learn how to manage their bodies in the physical world before they start spending large
amounts of time in the virtual world.

You can see children seeking out risks and thrills, together, in many
playground photos taken before the 1980s.[20] Some of them, such as figure
3.4, show playgrounds that are clearly *too* dangerous. If children fell from
such a great height, they could suffer severe injury, perhaps even a broken neck.

In contrast, figure 3.5 shows a playground spinner (or merry-go-round),
which is, in my opinion, the greatest piece of playground equipment ever
invented. It requires cooperation to get going: the more kids who join in,
the faster it goes and the more screaming there is, both of which amplify
the thrills. You get physical sensations from the centrifugal force that
you don't get anywhere else, which makes it educational as well as expe-
rientially unique. You get consciousness alteration if you lie in the center
(dizziness). To top it all off, it offers endless opportunities for additional
risk-taking such as standing up, hanging off the sides, or throwing a ball
with the other kids while it's spinning.

Figure 3.5. A playground spinner (or merry-go-round), a staple of 1970s playgrounds.[21]

On the playground spinner you can get hurt if you're not careful, but not badly hurt, which means you get direct feedback from your own skillful and unskillful moves. You learn how to handle your body and how to keep yourself and others safe. Researchers who study children at play have concluded that the risk of minor injuries should be a feature, not a bug, in playground design. In the U.K., they are acting on this insight, adding construction materials, hammers, and other tools (which are used with adult supervision).[22] As one enlightened summer camp administrator told me, "We want to see bruises, not scars."

Unfortunately, playground spinners are rare nowadays, because they carry *some* risk, and therefore in a litigious country like the United States they carry some risk of a lawsuit against whoever is responsible for the playground. You can see the decimation of risky play since the 1990s in most American playgrounds. Figure 3.6 shows the most common kind of structure in the playgrounds my children used in New York City in the early 2010s. It's hard to hurt yourself on these things, which means children don't learn much about how to *not* get hurt.

These ultrasafe structures were entertaining when my kids were

Figure 3.6. An overly safe playground, offering little opportunity for antifragile kids to learn how to not get hurt.[23]

Figure 3.7. Coney Island, New York City, offers a wide range of dosages of thrills.[24]

three or four, but by age 6 they wanted bigger thrills, which they found at Coney Island. Amusement parks around the world are designed to give children two of Sandseter and Kennair's six kinds of thrills: heights and high speeds. The rides offer differing doses of fear and thrill (with close to zero risk of injury), and a major topic of conversation in the car whenever I took my kids and their friends to Coney Island was, who is going to try which scary ride today?

Perhaps your first reaction to those old playground photos is "good riddance!" What parent wants to take *any* risks with their child? But the harms of eliminating all risky outdoor play are substantial. While writing this chapter, I met with Mariana Brussoni, a play researcher at the University of British Columbia. Brussoni guided me to research showing that the risk of injury per hour of physical play is *lower* than the risk per hour of playing adult-guided sports, while conferring many more developmental benefits (because the children must make all choices, set and enforce rules, and resolve all disputes).[25] Brussoni is on a campaign to encourage risky outdoor play because in the long run it produces the healthiest children.[26] Our goal in designing the places children play, she says, should be to "keep them as safe as necessary, not as safe as possible."[27]

The play researchers Brussoni, Sandseter and Kennair, and Peter Gray all help us see that antifragile children *need* play that involves some risk to develop competence and overcome their childhood anxieties. Like my dog, Wilma, only the kids themselves can calibrate the level of risk they are ready for at each moment as they tune up their experience-expectant brains. Like young trees exposed to wind, children who are routinely exposed to small risks grow up to become adults who can handle much larger risks without panicking. Conversely, children who are raised in a protected greenhouse sometimes become incapacitated by anxiety before they reach maturity.

I am often asked why I urge parents to be more vigilant and restrictive about their children's online activities when I've been talking for years about how parents need to stop over-supervising their children and start giving them independence. Can't children just as well become antifragile online? Don't they experience setbacks, stressors, and challenges there?

I see few indications that a phone-based childhood develops antifragility. Human childhood evolved in the real world, and children's minds are "expecting" the challenges of the real world, which is embodied, synchronous, and one-to-one or one-to-several, within communities that endure. For physical development they need physical play and physical risk-taking. Virtual battles in a video game confer little or no physical benefit. For social development they need to learn the art of friendship, which is embodied; friends do things together, and as children they touch, hug, and wrestle. Mistakes are low cost, and can be rectified in real time. Moreover, there are clear embodied signals of this rectification, such as an apology with an appropriate facial expression. A smile, a pat on the back, or a handshake shows everyone that it's okay, both parties are ready to move on and continue playing, both are developing their skills of relationship repair. In contrast, as young people move their social relationships online, those relationships become disembodied, asynchronous, and sometimes disposable. Even small mistakes can bring heavy costs in a viral world where content can live forever and everyone can see it. Mistakes can be met with intense criticism by multiple individuals with whom one has no underlying bond. Apologies are often mocked, and any signal of re-acceptance can be mixed or vague. Instead of gaining an experience of social mastery, a child is often left with a sense of social incompetence, loss of status, and anxiety about future social interactions.

This is why there is no contradiction when I say that parents should supervise less in the real world but more in the virtual—primarily by delaying immersion. Childhood evolved on Earth, and children's antifragility is geared toward the characteristics of Earth. Small mistakes promote growth and learning. But if you raise children on Mars, there's a mismatch between children's needs and what the environment offers. If a child falls down on Mars and cracks the face shield of their spacesuit, it's instant death. Mars is unforgiving, and life there would require living in defend mode. Of course, the online world is not nearly as dangerous as Mars, but it shares the property that small mistakes can bring enormous costs. Children did not evolve to handle the virality, anonymity, instability, and potential for large-scale public shaming of the virtual world. Even adults have trouble with it.

We are misallocating our protective efforts. We should be giving children more of the practice they need in the real world and delaying their entry into the online world, where the benefits are fewer and the guardrails nearly nonexistent.

THE BEGINNING OF THE END OF PLAY-BASED CHILDHOOD

At what age were you given freedom? How old were you when your parents let you walk alone to a friend's home, at least a quarter mile away, or allowed you and your friends to be out on your own, going to parks or shops, with no supervision? I have asked this question to dozens of audiences, and I always find the same generational differences.

First I ask everyone who was born before 1981 to raise their hands.

These are the members of Gen X (born 1965–1980), the baby boomers (born 1946–1964), and the last members of the so-called Silent Generation (born 1928–1945). I ask these older audience members to recall their age of liberation privately and then to shout it out when I point to their section of the room. Nearly everyone shouts out "6," "7," or "8," and it is sometimes hard for me to continue the demonstration because they are laughing and fondly recounting to each other the grand adventures they used to have with the other kids in their neighborhood. Next I ask everyone who was born in 1996 or later (Gen Z) to raise their hands. When I ask them to shout out their liberation age, the difference is stark: The majority fall between 10 and 12, with just a few 8s, 9s, 13s, and 14s. (Members of the millennial generation fall in between and show a wide range of liberation ages.)

These findings are confirmed by more rigorous research. In the United States,[28] Canada,[29] and Britain,[30] children used to have a great deal of freedom to walk to school, roam around their neighborhoods, invent games, get into conflicts, and resolve those conflicts, beginning around first or second grade. But in the 1990s, parenting changed in all three countries. It became more intensive, protective, and fearful.

Corresponding to the crackdown, studies of how Americans spend their time show a sudden change in the 1990s. Women had been entering the workforce in large numbers since the 1970s, giving them far less time at home. Yet despite growing time pressures, mothers as well as fathers began reporting that they spent a lot *more* time with their children, beginning rather suddenly in the mid-1990s. Figure 3.8 shows the changing number of hours per week that mothers reported spending with their children from 1965 through 2008. The number is steady or slightly declining, for mothers with and without college degrees, all the way until 1995, and then it jumps up, especially for college-educated mothers. The graph for fathers is quite similar, just with lower numbers (around four hours per week until 1995, then jumping up to around eight hours per week by 2000).

A separate study, looking at how children spend their time (as reported by parents), found that American children were also facing a time

Time Spent Parenting by Mothers

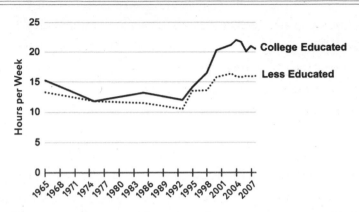

Figure 3.8. Time spent parenting by U.S. mothers. Parenting time suddenly increased in the mid-1990s—the beginning of Gen Z. (Source: Ramey & Ramey, 2000.)[31]

crunch that reduced their time for free play between 1981 and 1997.[32] Kids began spending more time in school and other structured (adult-supervised) activities and less time playing or watching TV. (The same thing happened in the U.K.[33]) What happened? Children were getting less time to play, but they suddenly got *more* time with their time-starved parents?

The authors of the study in figure 3.8 suggest that one contributing factor was the increasing focus in the 1990s on the competitiveness of college admissions. It's as if American parents, particularly in the top quarter of the income distribution, began thinking of their (fewer) children as precious and delicate race cars, and they—the parents—are the pit crew working frantically to help their car win the race to get into a top college.[34]

This theory fits with qualitative research done in the 1990s by the sociologist Annette Lareau, whose book *Unequal Childhoods*[35] chronicled the two basic parenting philosophies used by American parents. The first philosophy, which she calls "concerted cultivation," was the dominant model used by families in the middle and upper class. It begins with the premise that children require an extraordinary degree of care and training by adults. Parents must buy *Baby Einstein* videos to raise their children's IQs (even though researchers later showed such videos to be worthless[36]). Children's calendars must be filled with activities that the *parents* believe are enriching, such as learning Mandarin, or extra math training, even when such activities reduce autonomy and leave less room for free play.

Among the working class and the poor, Lareau found a very different approach, which she called "natural growth parenting." In this philosophy, kids will be kids, and if you just let them be, they'll turn into competent and responsible adults without a great deal of hand-holding. But surprisingly, a recent study of attitudes toward parenting found that by the 2010s many working-class parents had moved toward concerted cultivation parenting, including a high level of protection from risk.[37]

American parenting changed in the 1990s, first among college-educated parents and then more broadly. Fears of abduction and sex criminals had been rising since the 1980s, but even still, the general pattern for children in elementary and middle school up through the 1980s was that after school and on the weekends, kids were on their own to play in their neighborhoods in mixed-age groups, seek thrills, have adventures, work out conflicts, engage in antiphobic risk-taking, develop their intrinsic antifragility, enjoy being in discover mode together—and come home when the streetlights came on. Those after-school hours were probably more valuable for social development and mental health than anything that happened in school (other than recess).

FEARFUL PARENTING IN THE ANGLOSPHERE

The rapid loss of childhood autonomy was not, it turns out, primarily caused by parental fears about college admissions. That fear might have contributed to behavior change among middle- and upper-income Americans, but it can't explain why parents in Canada and Britain made the same changes at the same time in countries where college admissions are far less fraught. Psychologists and sociologists have pointed to several reasons why parents began giving their children less autonomy in the 1980s and 1990s, including gradual changes to urban design as cities and towns became more car-centric and urbanized. A related factor is a declining sense of social cohesion throughout the late 20th century, which had many causes. When people no longer knew their neighbors, they no longer had "eyes on the street" from adults who could look out for kids.[38] But perhaps the most important change in the 1980s was the rising fear among parents that everyone and everything was a threat to their children.[39]

In 2001, Frank Furedi, a British sociologist, published an important book titled *Paranoid Parenting: Why Ignoring the Experts May Be Best for Your Child*.[40] The book includes dozens of stories from the U.K. that sound as if they could have happened in the United States today, such as the mother who drove for hours, tailing the school bus that was taking her son on a class trip, to be sure the boy made it safely to the destination.

Furedi's book is of particular importance because it is written by an academic sociologist rather than by a parenting "expert." He analyzes changing parental behavior as a response to social, economic, and technological changes in the 1980s and 1990s: for example, the rise of cable TV (and 24/7 news cycles) and its ability to spread stories that frighten parents; the rising number of women working and the corresponding increase in day care and after-school programs; and the increasing influence of parenting "experts," whose advice was often a better reflection of their social and political views than of any scientific consensus.

Furedi says that there is one factor above all others that created the conditions for the 1990s turn to paranoid parenting: "the breakdown of adult solidarity." As Furedi explains,

Across cultures and throughout history, mothers and fathers have acted on the assumption that if their children got into trouble, other adults—often strangers—would help out. In many societies adults feel duty-bound to reprimand other people's children who misbehave in public.

But in Britain and America, the 1980s and 1990s saw repeated news stories about adults abusing children, from day care centers and sports leagues to the Boy Scouts and the Catholic Church. Some of these cases were true horror stories about institutions that had sheltered child abusers for decades in order to avoid bad publicity. Some of the cases were fabrications and moral panics[41]—in particular, those in which employees at day care centers were accused of carrying out bizarre sexual or satanic rituals. (The accusations were made by very young children, who, it later turned out, had invented imaginative stories in response to leading questions from overzealous adults.[42])

These scandals—real and fake—led to better detection and reporting mechanisms to catch abusers and hold institutions responsible for sheltering them. Their tragic side effect, however, was a generalized sense that no adults could be trusted to be alone with children. Children were taught to fear unknown adults, particularly men. According to Google's Ngram viewer (which charts the frequency of words and terms in all books published each year), the term "stranger danger" first appeared in English-language books in the early 1980s; then its frequency leveled off until the mid-1990s, after which it rose rapidly. At the same time, adults internalized the reciprocal message: Stay away from other people's children. Don't talk to them; don't discipline them if they are misbehaving; don't get involved.

But when adults step away and stop helping each other to raise children, parents find themselves on their own. Parenting becomes harder, more fear-ridden, and more time consuming, especially for women, as we saw in figure 3.8.

Furedi offered an important qualification about the scope of the problem: "The idea that responsible parenting means the continual supervision of children is a peculiarly Anglo-American one."[43] He noted that children in Europe, from Italy to Scandinavia, and in many other parts of the world, enjoyed far greater freedom to play and explore the outside world than did children in the U.K. and the United States. He cited a study showing that parents in Germany and Scandinavia were much more likely to let their young children walk to school than those in the U.K., who felt compelled to drive their children even short distances.[44]

It is this rise of fearful parenting in the 1990s that led to the evapo-ration of unsupervised children from public spaces in the Anglosphere by the year 2000. By almost any measure, children were safer in public than they had been in a very long time in terms of risks from crime, sex offenders, and even drunk drivers, all of which had been present at much higher levels in previous decades.[45] And once unsupervised children be-came a rarity, the occasional sighting of one was enough to cause some neighbors to call 911, bringing down the police, Child Protective Ser-vices, and occasionally jail time for anyone who dared to give their child the independence they themselves had enjoyed 30 years earlier.[46]

This is the world in which Gen Z was raised. It was a world in which adults, schools, and other institutions worked together to teach children that the world is dangerous, and to prevent them from experiencing the risks, conflicts, and thrills that their experience-expectant brains needed to overcome anxiety and set their default mental state to discover mode.[47]

SAFETYISM AND CONCEPT CREEP

The Australian psychologist Nick Haslam originated the term "concept creep,"[48] which refers to the expansion of psychological concepts in re-cent decades in two directions: downward (to apply to smaller or more trivial cases) and outward (to encompass new and conceptually unre-lated phenomena). You can see concept creep in action by observing the expansion of terms like "addiction," "trauma," "abuse," and "safety." For most of the 20th century, the word "safety" referred almost exclusively to physical safety. It was only in the late 1980s that the term "emotional safety" began to show up at more than trace levels in Google's Ngram viewer. From 1985 to 2010, at the start of the Great Rewiring, the term's frequency rose rapidly and steadily, a 600% increase.[49]

Physical safety is a good thing, of course. No sane person objects to the use of seat belts and smoke alarms. There is also an important con-cept called psychological safety, which refers to the shared belief in a group that members won't be punished or humiliated for speaking up, so people are willing to take risks in sharing ideas and debating them.[50]

*"We've created a safe, nonjudgmental environment that will
leave your child ill-prepared for real life.*

Figure 3.9. *New Yorker* cartoon by W. Haefeli.[51]

Psychological safety is among the best indicators of a healthy workplace culture. But in a psychologically safe group, members can disagree with each other and criticize each other's ideas respectfully. That's how ideas get vetted. What emerged on campus as emotional safety, in contrast, was a much broader concept that came to mean this: I should not have to experience negative emotions because of what someone else said or did. I have a right not to be "triggered."

In *The Coddling of the American Mind*, Greg and I found that the concept of safety had undergone such extensive concept creep among Gen Z and many of the educators and therapists around them that it had become a pervasive and unquestionable value. We used the term "safetyism" to refer to "a culture or belief system in which safety has become a sacred value, which means that people become unwilling to make trade-offs demanded by other practical and moral concerns. 'Safety' trumps everything else, no matter how unlikely or trivial the potential danger."[52] Students who had been raised with safetyism on the playground sometimes expected it to govern their classrooms, dorms, and campus events.

You can see the all-encompassing play-crushing power of safetyism in figure 3.10, sent to me by a friend in Berkeley, California. The administrators at this elementary school don't trust their students to play tag without adult guidance, because . . . what if there's a dispute? What if someone is excluded?

The school offers similarly inane lists of instructions and prohibitions to help children play other games. In the rules for playing touch football, the sign says FOOTBALL CAN ONLY BE PLAYED IF AN ADULT IS SUPERVISING AND REFEREEING THE GAME. The administrators seem to be committed to preventing the sorts of conflicts that are inherent in human interaction, and that would teach children how to manage their own affairs, resolve differences, and prepare for life in a democratic society.

American parents have lost so much trust in their fellow citizens and their own children that many now endorse the near-total elimination of freedom from childhood. According to a 2015 report from the Pew Re-

Figure 3.10. Restrictions on free play, at an elementary school in Berkeley, California.[53]

search Center, parents (on average) say children should be at least 10 years old to play unsupervised *in their own front yard.*[54] They say that kids should be at least 12 years old before being allowed to stay alone in their own home *unsupervised for one hour.* They say that kids should be 14 before being allowed to go, unsupervised, to *a public park.* And these respondents include the same Gen X and baby boom parents who say, gleefully and gratefully, that they were let out, in a much more dangerous era, at ages 6, 7, or 8.

ANTIFRAGILITY AND THE ATTACHMENT SYSTEM

Earlier in this chapter I described discover mode and defend mode as parts of a dynamic system for quickly adapting to changing conditions, like a thermostat. That system is embedded in a larger dynamic system called the attachment system. Mammals are defined by the evolutionary innovation that females bear their young live (not as eggs) and then produce milk to feed them. Mammal babies therefore have a long period of dependence and vulnerability during which they must achieve two goals: (1) develop competence in the skills needed for adulthood, and (2) don't get eaten. The best way to avoid getting eaten is generally to stick close to Mom. But as mammals mature, their experience-expectant brains need to wire up by practicing skills such as running, fighting, and befriending. This is why young mammals are so motivated to move away from Mom to play, including risky play.

The psychological system that manages these competing needs is called the attachment system. It was first described by the British psychoanalyst John Bowlby, who had studied the effects of separating children from their parents during World War II. Figure 3.11 is an excellent illustration of the attachment system in action, from the psychologist Deirdre Fay.

Every child needs at least one adult who serves as a "secure base." Usually it is the mother, but it can just as well be the father, grandparent, or nanny, or any adult who is reliably available for comfort and protection. If safety was the child's only goal, he'd stay "on base" for all of childhood.

The Attachment System

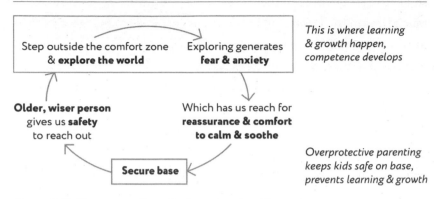

Figure 3.11. The mammalian attachment system.[55]

There'd be no need for a complicated regulatory system. But as soon as children can crawl, they want to crawl over to things they can touch, suck on, or otherwise explore. They need to spend a lot of time in discover mode, because that's where the learning and neural fine-tuning take place. But inevitably, something goes wrong. The child falls and bangs his head; a cat hisses at him; a stranger approaches. At that point defend mode activates and the child scurries back to base or starts crying, which is the child's way of calling for the base to come to him.

A securely attached child usually settles within a few seconds or minutes, shifts back to discover mode, and heads out for more learning. This process happens dozens of times a day, hundreds of times a month, and within a few years children become less fearful and more likely to want to explore on their own—perhaps by walking to school or a friend's house with no help from an adult.[56] As the child develops she is able to internalize the secure base. She doesn't need the parent's physical presence to feel that she has support, so she learns to face adversity by herself.

In adolescence, young people begin seeking out romantic relationships. These new attachments will reuse the psychological architecture and "internal working models" that were developed while forming attachments to parents. Adolescents will reuse those models to attach to love interests and later, perhaps, a spouse. But children who are kept on home base, prevented from making those off-base excursions that are so helpful for developing their antifragile nature, don't get to spend as much time in the growth zone. They may therefore spend more of their lives in defend mode, remaining more dependent on a parent's physical presence, which reinforces parental overprotection in a vicious cycle.

I have sketched out how things work in theory. In practice, everything about raising children is messy, hard to control, and harder to predict. Children raised in loving homes that support autonomy, play, and growth may still develop anxiety disorders; children raised in overprotective homes usually turn out fine. There is no one right way to be a parent; there is no blueprint for building a perfect child. Yet it is helpful to bear in mind some general features of human childhood: Kids are antifragile and therefore they benefit from risky play, along with a secure base, which helps to shift them over toward discover mode. A play-based childhood is more likely to do that than a phone-based childhood.

IN SUM

- The human brain contains two subsystems that put it into two common modes: *discover mode* (for approaching opportunities) and *defend mode* (for defending against threats). Young people born after 1995 are more likely to be stuck in defend mode, compared to those born earlier. They are on permanent alert for threats, rather than being hungry for new experiences. They are anxious.

- All children are by nature antifragile. Just as the immune system must be exposed to germs, and trees must be exposed to wind, children require exposure to setbacks, failures, shocks, and stumbles in order to develop strength and self-reliance. Overprotection interferes with this development and renders young people more likely to be fragile and fearful as adults.

- Kids must have a great deal of free play to develop, and they benefit from risky physical play, which has anti-phobic effects. Kids seek out the level of risk and thrill that they are ready for, in order to master their fears and develop competencies. Risk-taking online may not have comparable anti-phobic effects.

- In the 1980s and especially the 1990s, parents in Anglo countries became more fearful for many reasons, including changes in the media ecosystem and news cycle. They lost trust in each other, they started spending far more time supervising their own children, and they did more parenting in defend mode, seeing risks and threats everywhere.

- The worship of "safety" above all else is called safetyism. It is dangerous because it makes it harder for children to learn to care for themselves and to deal with risk, conflict, and frustration.

- The attachment system evolved to help young mammals learn the skills they'll need to reach adulthood while retreating to their "secure base" when they feel threatened. Fearful parenting keeps children on home base too much, preventing them from having the experiences they need to grow strong and to develop a secure attachment style.

- Children are most likely to thrive when they have a play-based childhood in the real world. They are less likely to thrive when fearful parenting and a phone-based childhood deprive them of opportunities for growth.

Chapter 4

PUBERTY AND THE BLOCKED
TRANSITION TO ADULTHOOD

From *The Ugly Duckling* to *The Very Hungry Caterpillar,* we reach for stories about animal metamorphosis to capture the emotions we feel as we watch our children grow and change. For humans, the change in bodies is not nearly as dramatic as it is for butterflies, yet the change in minds is every bit as extraordinary. But while caterpillars will turn into butterflies with little input from the outside world, the human transition from child to adult depends in part on getting the right kinds of experiences at the right time to guide the rapid rewiring of the adolescent brain.

PUBERTY, PLASTICITY, AND VULNERABILITY

As I noted in chapter 2, the human brain reaches 90% of its adult size by age 5, and it has far more neurons and synapses at that moment than it will have in its adult form. Subsequent brain development, therefore, is not about overall growth but about the selective pruning of neurons and synapses, leaving only the ones that have been frequently used. Brain researchers say, "Neurons that fire together, wire together,"[1] meaning that activities that repeatedly activate a constellation of neurons cause those neurons to connect more closely. If a child goes through puberty doing a lot of archery, or painting, or video games, or social media, those activities will cause lasting structural changes in the brain, especially if they are rewarding. This is how cultural experience changes the brain, producing a young adult who feels American instead of Japanese, or who is habitually in discover mode as opposed to defend mode.

A second kind of brain change that occurs during childhood is called myelination, which refers to the coating of the axons of neurons with an insulating sheath of a fatty material, which makes transmission faster across the long-distance connections in those constellations of neurons. These slow processes of pruning and myelination are related to the great trade-off of human brain development: The young child's brain has enormous *potential* (it can develop in many ways) but lower *ability* (it doesn't do most things as well as an adult brain). However, as pruning and myelination proceed, the child's brain becomes more efficient as it locks down into its adult configuration. This lockdown process happens in different parts of the brain at different times, and each lockdown is potentially the end of a sensitive period. It's like cement hardening: If you try to draw your name in very wet cement, it will disappear quickly. If you wait until the cement is dry, you'll leave no mark. But if you can catch it while it's in the transition between wet and dry, your name will last forever.[2]

Because pruning and myelination speed up at the start of puberty, changes in children's experiences during those years can have large and lasting effects.[3] In his textbook on adolescence, the developmental psychologist Laurence Steinberg notes that adolescence is not necessarily an especially *stressful* time. Rather, it is a time when the brain is *more vulnerable* to the effects of sustained stressors, which can tilt the adolescent into mental disorders such as generalized anxiety disorder, depression, eating disorders, and substance abuse. Steinberg adds,

Heightened susceptibility to stress in adolescence is a specific example of the fact that puberty makes the brain more malleable, or "plastic." This makes adolescence both a time of risk (because the brain's plasticity increases the chances that exposure to a stressful experience will cause harm) but also a window of opportunity for advancing adolescents' health and well-being (because the same brain plasticity makes adolescence a time when interventions to improve mental health may be more effective).[4]

Puberty is therefore a period when *we should be particularly concerned about what our children are experiencing*. Physical conditions, including nutrition, sleep, and exercise, matter throughout all of childhood and adolescence. But because there is a sensitive period for cultural learning, and because it coincides with the accelerated rewiring of the brain that begins at the start of puberty, those first few years of puberty deserve special attention.

EXPERIENCE BLOCKERS: SAFETYISM AND SMARTPHONES

Unlike carnivores, which evolved to get nearly all the nutrients they need by eating the flesh of other animals, humans are omnivores. We need to consume a wide variety of foods to get all the necessary vitamins, minerals, and phytochemicals. A child who eats only white foods (pasta, potatoes, chicken) will be nutrient deficient and at heightened risk of diseases such as scurvy (caused by a severe vitamin C deficiency).

Similarly, humans are socially and culturally adaptable creatures who need a wide variety of social experiences to develop into flexible and socially skilled adults. Because children are antifragile, it is essential that those experiences involve some fear, conflict, and exclusion (though not too much). Safetyism is an experience blocker. It prevents children from getting the quantity and variety of real-world experiences and challenges that they need.

How much stress and challenge does a child need in order to grow? Steinberg notes that "stressful experiences" are "ones that could cause harm." I wrote to him to ask if he agrees that children are antifragile and need exposure to short-term stressors—such as being excluded from a playgroup on one day—in order to develop resilience and emotional strength. He agreed that children are antifragile, and he added two qualifications to his statement about stressful experiences.

First, he noted that "chronic stress," meaning stress that lasts for days, weeks, or even years, is much worse than "acute stress," which refers to stress that comes on quickly but does not last long, such as an ordinary playground conflict. "Under chronic stress, it is much harder to adapt, recover, and get stronger from the challenge," he wrote. His second qualification was that "there is an inverted U-shaped pattern in the relationship between stress and well-being. A little stress is beneficial to development, but a lot of stress, acute or chronic, is detrimental."

Americans, Brits, and Canadians, unfortunately, tried to remove stressors and rough spots from children's lives beginning in the 1980s. Many parents and schools banned activities that they perceived as having *any* risk, not just of physical injury, but of emotional pain as well. Safetyism requires banning most independent activity during childhood, especially outdoor activities (such as playing touch football without an adult referee) because such activities could lead to bruised bodies and bruised feelings.

Safetyism was imposed on the millennials beginning slowly in the 1980s and then more quickly in the 1990s.[5] The rapid deterioration of mental health, however, did not begin until the early 2010s and was concentrated in Gen Z, not among millennials.[6] It was not until the addition of the second experience blocker—the smartphone—that rates began to rise.

Of course, using a smartphone *is* an experience. It is a portal to the infinite knowledge of Wikipedia, YouTube, and now ChatGPT. It connects young people to special-interest communities for everything from baking and books to extreme politics and anorexia. A smartphone makes it effortless for adolescents to stay in touch with dozens of people throughout the day and to join others in praising or shaming people.

In fact, smartphones and other digital devices bring so many interesting experiences to children and adolescents that they cause a serious problem: *They reduce interest in all non-screen-based forms of experience.* Smartphones are like the cuckoo bird, which lays its eggs in other birds' nests. The cuckoo egg hatches before the others, and the cuckoo hatchling promptly pushes the other eggs out of the nest in order to commandeer all of the food brought by the unsuspecting mother. Similarly, when a smartphone, tablet, or video game console lands in a child's life, it will push out most other activities, at least partially. The child will spend many hours each day sitting enthralled and motionless (except for one finger) while ignoring everything beyond the screen. (Of course, the same might be true of the parents as well, as the family sits "alone together.")

Are screen-based experiences less valuable than real-life flesh-and-blood experiences? When we're talking about children whose brains evolved to expect certain kinds of experiences at certain ages, yes. A resounding yes. Communicating by text supplemented by emojis is not going to develop the parts of the brain that are "expecting" to get tuned up during conversations supplemented by facial expressions, changing vocal tones, direct eye contact, and body language. We can't expect children and adolescents to develop adult-level real-world social skills when their social interactions are largely happening in the virtual world.[7] Synchronous video conversations are closer to real-life interactions but still lack the embodied experience.

If we want children to have a healthy pathway through puberty, we must first take them off experience blockers so that they can accumulate the wide range of experiences they need, including the real-world stressors their antifragile minds require to wire up properly. Then we should give children a clear pathway to adulthood with challenges, milestones, and a growing set of freedoms and responsibilities along the way.

RITES OF PASSAGE

On lists of human universals,[8] and on syllabi for introductory anthropology courses, you'll usually find rites of passage. This is because communities require rituals to signify shifts in people's status. It's the community's responsibility to conduct these rites, which commonly surround life events like birth (to welcome a new member and a new mother), marriage (to publicly declare a new social unit), and death (to acknowledge the departure of a member and the grieving of close kin). Most societies also have formal rites of passage around the time of puberty.

Despite the enormous variation in human cultures and gender roles, there is a common structure to puberty rites because they are all trying to do the same thing: Transform a girl into a woman or a boy into a man who has the knowledge, skills, virtues, and social standing to be an effective member of the community, soon to be ready for marriage and parenthood. In 1909, the Dutch-French ethnographer Arnold van Gennep noted that rites of passage around the world take the child through the same three phases. First, there is a *separation* phase in which young adolescents are removed from their parents and their childhood habits. Then there is a *transition* phase, led by adults other than the parents who guide the adolescent through challenges and sometimes ordeals. Finally, there is a *reincorporation* phase that is usually a joyous celebration by the community (including the parents), welcoming the adolescent as a new member of adult society, even though he or she will often receive years of further instruction and support.

Rites of passage for adolescents always reflect the structure and values of the adult society. Because all societies were highly gendered until recently, rites of passage have usually been different for girls and boys.

Rites for girls usually started soon after they had their first period, and the rites were often designed to prepare them for fertility and motherhood. Among Native Americans, for example, the Apache in Arizona still practice the "sunrise dance" after a girl's first period. The initiate is guided by an older woman (a sponsoring godmother chosen by the family for this honor) to build a temporary hut for herself, at some distance from the main campsite. These preparations are the separation phase, and they include bathing, hair washing, and the donning of new clothing, all of which emphasize purification and separation from all traces of childhood.[9]

The transition phase involves four days of highly prescribed dancing to rhythmic drumming and chanting from the older women. These dramatic ritual enactments are infused with a sense of sacredness. With the transition phase complete, the girl is welcomed joyously into womanhood, with feasting and exchanges of gifts between her family and others. She is reincorporated into her village and household, but now with new roles, responsibilities, and knowledge.

In traditional societies, the rites for turning a boy into a man are different from the rites for turning a girl into a woman. Because the visible signs of puberty are less obvious, the timing is more flexible. In many societies, boys are initiated as a group—all the boys around a certain age, who will become tightly bonded to each other by their shared ordeal. In societies that experienced frequent armed conflict with neighboring groups, a warrior ethos usually developed among men, and the transition phase often included a requirement to undergo physical pain, including body piercings or circumcision, to test and then publicly validate one's manhood. In many Indigenous North American societies, such as the Blackfoot in the Great Plains, the transition phase involved a vision quest in which the boy had to go out alone to a sacred site, chosen by the elders, where he fasted for four days while praying to the spirits for a vision or revelation of his purpose in life and the role he was to play in his community.[10]

Societies that were not preparing boys for war had very different kinds of rites for boys. In all Jewish communities, boys become subject to the laws of the Torah at the age of 13, and among their main duties as Jewish men, traditionally, was the study of the Torah. The Jewish rite of passage—the Bar Mitzvah—therefore involves a long period of instruction by a rabbi or scholar (not the father), followed by the big day when the boy takes the place of the rabbi in Saturday Shabbat services and reads the weekly Torah and haftorah portions in Hebrew.[11] In some Jewish communities, the boy then also delivers a commentary on what he has read. It is a challenging public performance for a boy who usually still looks like a child.

In Judaism, girls become subject to the commandments at the age of

12, which is likely an ancient recognition that girls enter puberty a year or two before boys. All but the most tradition-oriented congregations perform a ceremony for girls identical to the Bar Mitzvah called a Bat Mitzvah (which means "daughter of the commandment"; "bar" means son). Just because rites of passage always used to be gendered does not mean that they must be gendered today.

The fact that most societies used to have such rites suggests to me that our extremely new secular societies may be losing something important as we abandon public and communally marked rites of passage. A human child doesn't morph into a culturally functional adult solely through biological maturation. Children benefit from role models (for cultural learning), challenges (to stimulate antifragility), public recognition of each new status (to change their social identity), and mentors who are not their parents as they mature into competent, flourishing adults. Evidence for the idea that children *need* rites of passage comes from the many cases where adolescents spontaneously construct initiation rites that are not supported by adults in the broader culture. In fact, anthropologists say that such rites come about precisely because of a society's "failure to provide meaningful adolescent rites of passage ceremonies."[12]

Such constructions are perhaps most vivid among groups of boys, especially when those boys must bond together to be more effective in competing with other groups of boys. Think of the initiation rites that young men develop for entry into a college fraternity, secret society, or street gang.[13] When boys and young men have the freedom to create their own rituals, it often looks as though at least one of them took that intro to anthropology class. They spontaneously create rituals of separation, transition, and incorporation (into peer groups) that we outsiders lump together as "hazing." But because boys and young men construct these rites with little or no guidance from elders, the rituals can become cruel and dangerous. The resulting culture can also be dangerous for women when these young men try to demonstrate their manhood to their brothers in ways that exploit or humiliate women.

Girls construct rites of passage too, as when a college sorority inducts new members. These rites tend not to include as much physical pain as is demanded by boys, but there is often psychological pain related to beauty and sexuality. Initiates describe being rated, compared, and shamed for their physical features.[14]

Despite the pain and humiliation required for entry, many young people are willing to participate in these rites for the opportunity to join a binding social group and to transition away from childhood's parental dependency and into peer-oriented young adulthood. This suggests that there may be a deep need among many adolescents for belonging and for the rites and rituals that create and express that belonging. Can we use that knowledge to improve adolescents' transitions to adulthood?

WHY DO WE BLOCK THE TRANSITION TO ADULTHOOD?

I used the metaphor that puberty is like the chrysalis stage of a butterfly's life. But whereas the caterpillar hides away to emerge a few weeks later as a butterfly, the human child must undergo the transition in public over several years. Historically, there were plenty of adults, norms, and rituals to help the child along. But since the early 20th century, scholars have noted the disappearance of adolescent rites of passage across modern industrial societies. Such rites are now mostly confined to religious traditions, such as the Bar and Bat Mitzvah for Jews, the quinceañera celebration of a girl's 15th birthday among Catholic Latin Americans, and confirmation ceremonies for teens in many Christian denominations. These remaining rites are likely to be less transformative than they once were as religious communities become less central to children's lives in recent decades.[15]

Even without formal initiation rites, modern secular societies retained a few developmental milestones until quite recently. Those of us who grew up in the analog world of 20th-century America can remember a time when there were three nationally recognized age transitions that granted greater freedom and called for greater maturity:

- At 13 you were thought to be mature enough to go to a movie theater without a parent because most of the movies you wanted to see were rated PG-13.

- At 16 you could begin to drive (in most states). Cars were quasi-sacred objects for American teens, so this was a major birthday after which a new world of independent experience opened up for you. You had to learn to drive responsibly in the eyes of the state and of your parents or you'd lose this privilege.

- At 18 you were considered an adult. You could legally enter a bar or buy alcohol in a liquor store.[16] You could buy cigarettes in most states (although that varied). You could vote, and if you were male, you had to register for potential military service. Also, high school graduation occurs around age 18, and that was the end of formal education for many people. After high school, graduates were expected to get a job or head off to college. In either case, there would be a major break with childhood and a big step toward adulthood.

In the real world, it often matters how old you are. But as life moved online, it mattered less and less. The mass movement from real world to virtual world started with the rise of fearful parenting and the gradual loss of play-based childhood. As overprotection and safetyism intensified in the 1990s, young people began engaging less in some of the major activities traditionally associated with teen development, activities that often required a car and permission to be out of the house, unsupervised.

Figure 4.1 shows the percentage of U.S. high school seniors (roughly age 18) who had gotten a driver's license, and who had ever drunk alcohol, worked for pay, or had sexual intercourse. As you can see, the downturn in these activities did *not* start in the early 2010s; it started back in the 1990s and early 2000s.

At the same time that adults were reducing young people's access to the real world, the virtual world was becoming more accessible and more enticing. In the 1990s, millennial adolescents started spending more time on home computers connected to the internet. Computers became

Teens Engaging in Adult Activities

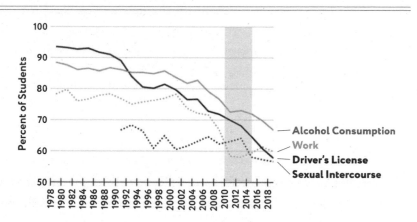

Figure 4.1. The percentage of U.S. high school seniors who have engaged in four adult activities has been declining since the 1990s or early 2000s, prior to the Great Rewiring of 2010 to 2015. (Source: Monitoring the Future and CDC Youth Risk Behavior Survey.)[17]

portable (laptops) and faster (higher connection speeds). But in the new virtual world, it almost never mattered how old you were. As soon as children could use a web browser, they had virtually unlimited access to everything on the web. And once teens moved from basic phones to smartphones, in the early 2010s, they could experience everything all day long. There is no equivalent to movie ratings such as PG-13, R, and X in the online world. Social media platforms such as Instagram, Snapchat, and TikTok don't enforce their minimum age of 13.[18] Children are free to do as they please, and to play video games and exchange messages and photographs with unknown adults. Pornography sites also welcome children, as long as they click a box to say that they are 18 or older. Porn sites will show them how to have anal sex long before they've had their first kiss.

Once a child gets online, there is never a threshold age at which she is granted more autonomy or more rights. On the internet, everyone is the same age, which is no particular age. This is a major reason why a phone-based adolescence is badly mismatched with the needs of adolescents.

In short, adults are doing a variety of things to Gen Z, often with good intentions, that prevent adolescents from experiencing a widely shared and socially validated progression from childhood dependency to adult independence. We interfered with their growth in the 1980s and 1990s when we blocked them from risky play and ramped up adult supervision and monitoring. We gave them unfettered access to the internet instead, removing all age thresholds that used to mark the path to adulthood. A few years later, we gave their younger siblings smartphones in middle school. Once we had a new generation hooked on smartphones (and other screens) *before* the start of puberty, there was little space left in the stream of information entering their eyes and ears for guidance from mentors in their real-world communities *during* puberty. There was just an infinite river of digital experience, customized for each child to maximize clicks and ad revenue, to be consumed alone in his or her room. It all got worse during the COVID pandemic years of "social distancing" and online everything.

But it doesn't have to be this way.

BUILDING A LADDER FROM CHILDHOOD TO ADULTHOOD

A country that is large, secular, and diverse by race, religion, and politics may not be able to construct shared rites of passage that are full of moral guidance, like the Apache sunrise ceremony. Yet despite our differences, we all want our children to become socially competent and mentally healthy adults who are able to manage their own affairs, earn a living, and form stable romantic bonds. If we can agree on that much, then might we be able to agree on norms that lay out some of the steps on that path? Importantly, these would mostly be norms, not laws, which any parent could choose to follow or ignore. Engaging in commonly held norm-based rituals and sharing milestones might be more effective than practices that each family invents for itself.

As an initial proposal, to start a conversation, I suggest that we focus on *even-year birthdays* from ages 6 to 18. We might make a big deal out of those birthdays by linking them to new freedoms, new responsibilities, and significant increases in allowance. We want children to feel that they are climbing a ladder with clearly labeled rungs, rather than just having an annual party with games, cake, and presents. It might look something like this:

Age 6: The age of family responsibility. Children are formally recognized as important contributors to the household, not just as dependents. As an example, they can be given a small list of chores and a small weekly allowance that is contingent upon their performance of those chores.[19]

Age 8: The age of local freedom. Children gain the freedom to play and hang out in groups without adult supervision. They should show that they can take care of each other, and they begin running local errands, if there are stores within a short walk or bike ride. They should not be given adult cell phones, but they could be given a phone or watch designed for children that would allow them to call or text a small number of people (such as their parents and siblings).

Age 10: The age of roaming. Preteens gain the freedom to roam more widely, perhaps equivalent to what their parents were allowed to do at the age of 8 or 9. They should show good judgment and do more to help their families. Consistent with their increased mobility and responsibility, a flip phone or other basic phone with few apps and no internet access might be given as a birthday present. They should not have most afternoons filled with adult-led "enrichment" activities; they need time to hang out with friends in person.

Age 12: The age of apprenticeship. At 12, which is around the age that many societies begin rites of initiation, adolescents should begin finding more adult mentors and role models beyond their parents. Adolescents should be encouraged to start earning their own money by doing chores for neighbors or relatives, such as raking leaves or working as a mother's helper for a neighbor with an infant or toddler. They might be encouraged to spend more time with trusted relatives, without their parents present.

Age 14: The beginning of high school. The 14th birthday comes around the time that high school begins, and this is a major transition during which independence increases along with academic pressure, time pressure, and social pressure. Activities such as working for pay and joining an athletic team are good ways to discover that hard work leads to tangible and pleasurable rewards. The beginning of high school would be a reasonable target for a national norm (not a law) about the minimum age at which teens get their first smartphone.[20]

Age 16: The beginning of internet adulthood. This should be a big year of independence, conditional on showing a history of responsibility and growth since the previous step. The U.S. Congress should undo the mistake it made in 1998 when it made 13 the age at which children can sign contracts with corporations to open accounts and give away their data without their parents' knowledge or consent. I believe the age should be raised to 16 and enforced. The 16th birthday would become a major milestone at which we say to teens, "You can now get a driver's license, and you can now sign certain kinds of contracts without any legal requirement for parental consent. You can now open social media accounts as well." (There are good arguments for waiting until 18, but I think 16 would be the right *minimum* age to be established by law.)

Age 18: The beginning of legal adulthood. This birthday would retain all of its legal significance including the beginning of voting, eligibility for military service, and the ability to sign contracts and make life decisions. Because this birthday falls near high school graduation in the United States, it should be treated in van Gennep's terms as both a *separation* from childhood and the beginning of a *transition* period into the next phase of life.

Age 21: Full legal adulthood. This birthday is the last one with any legal significance in the United States and many countries. At this age one can buy alcohol and cigarettes. One can enter casinos and sign up for internet sports gambling. The person is now a full adult in the eyes of the law.

These are my suggestions for a path to adulthood in a modern secular society. Your environment may be different, and your child may need to move along a different path at a different speed. But we should not let our variations force us to remove *all* common milestones and leave children to just wander around without any shared standards or age-graded increases in freedoms and responsibilities. Children do not turn into fully functioning adults on their own. Let's lay out some steps they can take that will help them to get there.

IN SUM

- Early puberty is a period of rapid brain rewiring, second only to the first few years of life. Neural pruning and myelination are occurring at a very rapid rate, guided by the adolescent's experiences. We should be concerned about those experiences and not let strangers and algorithms choose them.

- Safetyism is an experience blocker. When we make children's safety a quasi-sacred value and don't allow them to take any risks, we block them from overcoming anxiety, learning to manage risk, and learning to be self-governing, all of which are essential for becoming healthy and competent adults.

- Smartphones are a second kind of experience blocker. Once they enter a child's life, they push out or reduce all other forms of non-phone-based experience, which is the kind that their experience-expectant brains most need.

- Rites of passage are the curated sets of experiences that human societies arrange to help adolescents make the transition to adulthood. Van Gennep noted that these rites usually have a separation phase, a transformation phase, and a reincorporation phase.

- Western societies have eliminated many rites of passage, and the digital world that opened up in the 1990s eventually buried most milestones and obscured the path to adulthood. Once children began spending much or most of their time online, the inputs to their developing brains became undifferentiated torrents of stimuli with no age grading or age restrictions.

- A society that is large, diverse, and secular (such as the United States or the U.K.) might still agree to a set of milestones that mark stepwise increases in freedoms and responsibilities.

THIS CONCLUDES PART 2, WHICH PRESENTED THE LEAD-UP TO THE GREAT Rewiring of Childhood that occurred between 2010 and 2015. I explained why human childhood has the unique features that it has and why a play-based childhood is so well matched to those features. I showed evidence that the play-based childhood was in retreat well before the arrival of smartphones. Now we're ready to move on to part 3, in which I'll tell the story of what happened when adolescents transitioned from basic phones to smartphones, which began in the late 2000s and accelerated in the early 2010s. I'll present the evidence that the new phone-based childhood that emerged in those years is bad for children and adolescents, and I'll show that the harm goes far beyond increases in mental illness.

Part 3

THE GREAT REWIRING

THE RISE OF THE

PHONE-BASED CHILDHOOD

Chapter 5

THE FOUR FOUNDATIONAL HARMS: SOCIAL DEPRIVATION, SLEEP DEPRIVATION, ATTENTION FRAGMENTATION, AND ADDICTION

One morning, on a family trip to Vermont in 2016, my six-year-old daughter was playing a video game on my iPad. She called out to me: "Daddy, can you take the iPad away from me? I'm trying to take my eyes off it but I can't." My daughter was in the grip of a *variable-ratio reinforcement schedule* administered by the game designers, which is the most powerful way to take control of an animal's behavior short of implanting electrodes in its brain.

In 1911, in one of the foundational experiments in psychology, Edward Thorndike put hungry cats into "puzzle boxes." These were small cages from which the cat could escape and get food if it performed a particular behavior, such as pulling on a ring connected to a chain that opened the latch. The cats thrashed around unhappily, trying to escape, and they hit on the solution eventually. But what do you think happened the next time the same cat was put into that same box? Did it go right for the ring? No. Thorndike found that the cats thrashed around again, although on average they hit upon the solution a bit faster the second time, and a bit faster each time after that, until they performed the escape behavior immediately. There was always a learning curve. There was never a moment of insight in which the cat "got it" and the times suddenly dropped.

Thorndike described the cat's learning like this: "The one impulse, out of many accidental ones, which leads to pleasure, becomes strengthened and stamped in." He said that animal learning is "the wearing smooth of a path in the brain, not the decisions of a rational consciousness."[1] Keep that phrase in mind whenever you see anyone (including yourself) making repetitive motions on a touch screen, as if in a trance: "the wearing smooth of a path in the brain."

MY GOAL IN PART 3 IS TO EXAMINE THE EVIDENCE OF HARM FROM THE Great Rewiring across a wide spectrum of outcomes. The rapid switch from flip phones (and other basic phones) to smartphones with high-speed internet and social media apps created the new phone-based childhood, which laid down many new paths in the brains of Gen Z. In this chapter, I describe the four foundational harms of the new phone-based childhood that damage boys and girls of all ages: social deprivation, sleep deprivation, attention fragmentation, and addiction. Then, in chapter 6, I'll lay out the main reasons why social media has been especially damaging to girls, including chronic social comparison and relational aggression. In chapter 7, I'll examine what's going wrong for boys, whose mental health did not decline as suddenly as it did for girls, but who have been withdrawing from the real world and investing ever more of their efforts in the virtual world for several decades. In chapter 8, I'll show that the Great Rewiring encouraged habits that are exactly contrary to the accumulated wisdom of the world's religious and philosophical traditions. I'll show how we can draw on ancient spiritual practices for guidance on how to live in our confusing, overwhelming time. But first, I need to explain what the phone-based childhood is and where it came from.

THE ARRIVAL OF THE PHONE-BASED CHILDHOOD

When Steve Jobs announced the first iPhone in June 2007, he described it as "a widescreen iPod with touch controls, a revolutionary mobile phone, and breakthrough internet communication device."[2] The first iteration of the iPhone was quite simple by today's standards, and I have no reason to believe it was harmful to mental health. I bought one in 2008 and found it to be a remarkable digital Swiss Army knife, full of tools I could call on when I needed them. It even had a flashlight! It was not designed to be addictive or to monopolize my attention.

This soon changed with the introduction of software development kits, which allowed third-party apps to be downloaded onto mobile devices. This revolutionary move culminated in the launch of the App Store by Apple in July 2008, starting with 500 apps available. Google followed suit with the Android Market in October 2008, which was rebranded and expanded into Google Play in 2012. By September 2008, the Apple App Store had grown to hold more than 3,000 apps, and by 2013 it had more than 1 million.[3] The Google Play store grew right alongside Apple, reaching 1 million apps in 2013.[4]

The opening of smartphones to third-party apps led to fierce competition among companies large and small to create the most engaging mobile apps. The winners of this race were often those that adopted free-to-use, advertising-based business models because few consumers would pay $2.99 for an app if a competitor offered one for free. This proliferation of advertising-driven apps caused a change in the nature of time spent using a smartphone. By the early 2010s, our phones had transformed from Swiss Army knives, which we pulled out when we needed a tool, to platforms upon which companies competed to see who could hold on to eyeballs the longest.[5]

The people with the least willpower and the greatest vulnerability to manipulation were, of course, children and adolescents, whose frontal cortices were still highly underdeveloped. Children have been drawn powerfully to screens since the advent of television, but they could not take those screens with them to school or when they went outside to play. Before the iPhone, there was a limit to the amount of screen time a child could have, so there was still time for play and face-to-face conversation. But the explosion of smartphone-based apps such as Instagram in the exact years in which teens and preteens were moving from basic phones to smartphones marked a qualitative change in the nature of childhood. By 2015, more than 70% of American teens carried a touch screen around with them,[6] and these screens became much better at holding their attention, even when they were with their friends. This is why I date the beginning of the phone-based childhood to the early 2010s.

As I noted in the introduction, I use the term "phone-based" in an expansive sense to include *all internet-connected devices*. In the late 2000s and early 2010s, many of these devices, particularly video game consoles such as the PS3 and Xbox 360, gained access to the internet, introducing advertising and new commercial incentives to platforms that had once been self-enclosed. Insofar as laptops with high-speed internet provided access to social media platforms, internet-based computer games, and free streaming platforms with user-generated videos (including YouTube and many online pornography sites), they are part of the phone-based childhood too. I use the term "childhood" here expansively as well, to include both childhood and adolescence.

SOCIAL MEDIA AND ITS TRANSFORMATIONS

Social media has evolved over time,[7] but there are at least four major features common to the platforms we generally think of as being clear examples of social media: *user profiles* (users can create individual profiles where they can share personal information and interests); *user-generated content* (users create and share a variety of content to a broad audience, including text posts, photos, videos, and links); *networking* (users can connect with other users by following their profiles, becoming friends, or joining the same groups); and *interactivity* (users interact with each other and with the content they share; interactions may include liking, commenting, sharing, or direct messaging). The prototypical social media platforms such as Facebook, Instagram, Twitter, Snapchat, TikTok, Reddit, and LinkedIn share all four features, as does YouTube (even though YouTube is more widely used as the world's video library than for its social features) and also the now popular video game streaming platform Twitch. Even modern adult-content sites, like OnlyFans, have adopted these four features. On the other hand, messaging apps such as WhatsApp and Facebook Messenger do not have all four features, and while they are certainly social, they would not be considered social media.

A transformational shift in the nature of social media happened in the years around 2010 that made it more harmful to young people. In the early years of Facebook, Myspace, and Friendster (all founded between 2002 and 2004), we called these services *social networking systems* because they were primarily about connecting individuals, such as long-lost high school friends or fans of a particular musician. But around 2010 there was a series of innovations that fundamentally changed these services.

First and foremost, in 2009, Facebook introduced the "like" button and Twitter introduced the "retweet" button. Both of these innovations were then widely copied by other platforms, making viral content dissemination possible. These innovations quantified the success of every post and incentivized users to craft each post for maximum spread, which sometimes meant making more extreme statements or expressing more anger and disgust.[8] At the same time, Facebook began using algorithmically curated news feeds, which motivated other platforms to join the race and curate content that would most successfully hook users. Push notifications were released in 2009, pinging users with notifications throughout the day. The app store brought new advertising-driven platforms to smartphones. Front-facing cameras (2010) made it easier to take photos and videos of oneself, and the rapid spread of high-speed internet (reaching 61% of American homes by January 2010[9]) made it easier for everyone to consume everything quickly.

By the early 2010s, social "networking" systems that had been structured (for the most part) to connect people turned into social media "platforms" redesigned (for the most part) in such a way that they encouraged one-to-many public performances in search of validation, not just from friends but from strangers. Even users who don't actively post are affected by the incentive structures these apps have designed.[10]

These changes explain why the Great Rewiring began around 2010 and why it was largely complete by 2015. Children and adolescents, who were increasingly kept at home and isolated by the national mania for overprotection, found it ever easier to turn to their growing collection of internet-enabled devices, and those devices offered ever more attractive and varied rewards. The play-based childhood was over; the phone-based childhood had begun.

THE OPPORTUNITY COST OF A PHONE-BASED CHILDHOOD

Suppose a salesman in an electronics store told you he had a new product for your 11-year-old daughter that would be very entertaining—even more than television—with no harmful side effects of any kind, but also no more than minimal benefits beyond the entertainment value. How much would this product be worth to you?

You can't answer this question without knowing the *opportunity cost*. Economists define that term as the loss of other potential gains when one alternative is chosen. Suppose you are starting a business and you consider paying $2,000 to take a course on graphic design at a local university so that you can make your company's communications look better. You can't just ask yourself whether more attractive flyers and websites will earn back the $2,000. You have to consider all the other things you could have done with that money—and, perhaps more importantly, what else you could have done to help your business with all the *time* you spent taking the course.

So, when that salesman tells you that the product is free, you ask about the opportunity cost. How much time does the average child spend using the product? Around 40 hours a week for preteens like your daughter, he says. For teens aged 13 to 18, it's closer to 50 hours per week. At that point, wouldn't you walk out of the store?

Those numbers—six to eight hours per day—are what teens spend on all screen-based leisure activities.[11] Of course, children were already spending a lot of their time watching TV and playing video games before the smartphone and internet became parts of their daily lives. Long-running studies of American adolescents show that the average teen was watching a little less than three hours per day of television in the early 1990s.[12] As most families gained dial-up access to the internet during that decade, followed by high-speed internet in the 2000s, the amount of time spent on internet-based activities increased, while time spent watching TV decreased. Kids also began to spend more time playing video games and less time reading books and magazines. Putting it all together, the Great Rewiring and the dawn of the phone-based childhood seem to have added two to three hours of *additional* screen-based activity, on average, to a child's day, compared with life before the smartphone. These numbers vary somewhat by social class (more use in lower-income families than in high-income families), race (more use in Black and Latino families than in white and Asian families[13]), and sexual minority status (more use among LGBTQ youth; see more detail in this endnote[14]).

I should note that researchers' efforts to measure screen time are probably underestimates. When the question is asked differently, Pew Research finds that a third of teens say they are on one of the major social media sites "almost constantly,"[15] and 45% of teens report that they use the internet "almost constantly." So even if the average teen reports "just" seven hours of leisure screen time per day, if you count all the time that they are actively *thinking* about social media while multitasking in the real world, you can understand why nearly half of all teens say that they are online almost all the time. That means around 16 hours per day—112 hours per week—in which they are not fully present in whatever is going on around them. This kind of continuous use, often involving two or three screens at the same time, was simply not possible before kids carried touch screens in their pockets. It has enormous implications for cognition, addiction, and the wearing smooth of paths in the brain, especially during the sensitive period of puberty.

In *Walden*, his 1854 reflection on simple living, Henry David Thoreau wrote, "The cost of a thing is the amount of . . . life which is required to be exchanged for it, immediately or in the long run."[16] So what was the opportunity cost to children and adolescents when they started spending six, or eight, or perhaps even 16 hours each day interacting with their devices? Might they have exchanged any parts of life that were necessary for healthy human development?

HARM #1: SOCIAL DEPRIVATION

Children need a lot of time to play with each other, face to face, to foster social development.[17] But back in chapter 2, I showed that the percentage of 12th graders who said that they got together with their friends "almost every day" dropped sharply after 2009.

You can see the loss of friend time in finer detail in figure 5.1, from a study on how Americans of all ages spend their time.[18] The figure shows the daily average number of minutes that people in different age brackets spend with their friends. Not surprisingly, the youngest group (ages 15–24) spends more time with friends, compared with the older groups, who are more likely to be employed and married. The difference was very large in the early 2000s, but it was declining, and the decline accelerated after 2013. The data for 2020 was collected after the COVID epidemic arrived, which explains why the lines bend downward in that last year for the two older groups. But for the youngest age group there is no bend at 2019. The decline caused by the first year of COVID restrictions was no bigger than the decline that occurred the year before COVID arrived. In 2020, we began telling everyone to avoid proximity to any person outside their "bubble," but members of Gen Z began socially distancing themselves as soon as they got their first smartphones.

Of course, teens at the time might not have thought they were losing their friends; they thought they were just moving the friendship from real life to Instagram, Snapchat, and online video games. Isn't that just

Daily Time with Friends, by Age Group

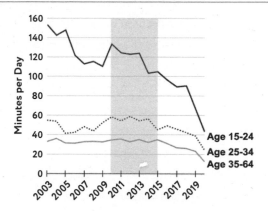

Figure 5.1. Daily average time spent with friends in minutes. Only the youngest age group shows a sharp drop before the 2020 data collection, which was performed after COVID restrictions had begun. (Source: American Time Use Study.)[19]

as good? No. As Jean Twenge has shown, teens who spend more time using social media are more likely to suffer from depression, anxiety, and other disorders, while teens who spend more time with groups of young people (such as playing team sports or participating in religious communities) have better mental health.[20]

It makes sense. Children need face-to-face, synchronous, embodied, physical play. The healthiest play is outdoors and includes occasional physical risk-taking and thrilling adventure. Talking on FaceTime with close friends is good, like an old-fashioned phone call to which a visual channel has been added. In contrast, sitting alone in your bedroom consuming a bottomless feed of other people's content, or playing endless hours of video games with a shifting cast of friends and strangers, or posting your own content and waiting for other kids (or strangers) to like or comment is so far from what children need that these activities should not be considered healthy new forms of adolescent interaction; they are alternatives that consume so much time that they reduce the amount of time teens spend together.

The sharp drop of time with friends actually *underestimates* the social deprivation caused by the Great Rewiring because even when teens are within a few feet of their friends, their phone-based childhoods damage the quality of their time together. Smartphones grab our attention so powerfully that if they merely vibrate in our pockets for a tenth of a second, many of us will interrupt a face-to-face conversation, just in case the phone is bringing us an important update. We usually don't tell the other person to stop talking; we just pull out our phone and spend some time pecking at it, leaving the other person to conclude, reasonably, that she is less important than the latest notification. When a conversation partner pulls out a phone,[21] or when a phone is merely visible[22] (not even your own phone), the quality and intimacy of a social interaction is reduced. As screen-based technologies move out of our pockets and onto our wrists, and into headsets and goggles, our ability to pay full attention to others is likely to deteriorate further.

It's painful to be ignored, at any age. Just imagine being a teen trying to develop a sense of who you are and where you fit, while everyone you meet tells you, indirectly: You're not as important as the people on my phone. And now imagine being a young child. A 2014 survey of children ages 6–12, conducted by *Highlights* magazine, found that 62% of children reported that their parents were "often distracted" when the child tried to talk with them.[23] When they were asked the reasons why their parents were distracted, cell phones were the top response. Parents know that they are shortchanging their own children. A 2020 Pew survey found that 68% of parents said that they sometimes or often feel distracted by their phones when they are spending time with their children. Those numbers were higher for parents who were younger and who were college educated.[24]

The Great Rewiring devastated the social lives of Gen Z by connecting them to everyone in the world and disconnecting them from the people around them. As a Canadian college student wrote to me,

> Gen Z are an incredibly isolated group of people. We have shallow friendships and superfluous romantic relationships that are mediated and governed to a large degree by social media. . . . There is hardly a sense of community on campus and it's not hard to see.

Oftentimes I'll arrive early to a lecture to find a room of 30+ students sitting together in complete silence, absorbed in their smartphones, afraid to speak and be heard by their peers. This leads to further isolation and a weakening of self identity and confidence, something I know because I've experienced it firsthand.[25]

HARM #2: SLEEP DEPRIVATION

Parents have long struggled to get their children to go to bed on school nights, and smartphones have exacerbated this struggle. Natural sleep patterns shift during puberty.[26] Teens start to go to bed later, but because their weekday mornings are dictated by school start times, they can't sleep later. Rather, most teens just get less sleep than their brains and bodies need. This is a shame because sleep is vital for good performance in school and life, particularly during puberty, when the brain is rewiring itself even faster than it did in the years before puberty. Sleep-deprived teens cannot concentrate, focus, or remember as well as teens who get sufficient sleep.[27] Their learning and their grades suffer.[28] Their reaction times, decision making, and motor skills suffer, which elevates their risk of accidents.[29] They are more irritable and anxious throughout the day, so their relationships suffer. If sleep deprivation goes on long enough, other physiological systems become perturbed, leading to weight gain, immune suppression, and other health problems.[30]

Teens need more sleep than adults—at least nine hours a night for preteens and eight hours a night for teens.[31] Back in 2001, a leading sleep expert wrote that "almost all teenagers, as they reach puberty, become walking zombies because they are getting far too little sleep."[32] When he wrote that, sleep deprivation had been rising for a decade, as you can see in Figure 5.2. Sleep deprivation then leveled off through the early 2010s. After 2013, it resumed its upward march.

Is that just a coincidence, or is there evidence directly linking the upsurge in sleep problems to the arrival of the phone-based childhood? There's a lot of evidence. A review of 36 correlational studies found significant associations between high social media use and poor sleep, and

Teens Who Get Less Than 7 Hours of Sleep

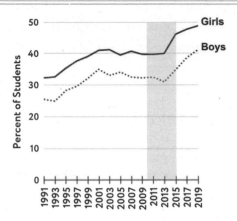

Figure 5.2. Percent of U.S. students (8th, 10th, and 12th grade) who get less than seven hours of sleep on most nights. (Source: Monitoring the Future.)[33]

also between high social media use and poor mental health outcomes.[34] That same review also found that high social media use at one time predicted sleep problems and worse mental health at later times. One experiment found that adolescents who restricted their use of screen devices after 9 p.m. on school nights for two weeks showed increased total sleep time, earlier sleep onset times, and improved performance on a task that required focused attention and quick reactions.[35] Other experiments, using a variety of different screen-based technologies (including e-readers, video games, and computers), have also found that late-night use is disruptive to sleep.[36] Thus, the relationships are not merely correlations; they are causal.

It makes intuitive sense. A study by Jean Twenge and colleagues of a large U.K. data set found that "heavy use of screen media was associated with shorter sleep duration, longer sleep latency, and more mid-sleep awakenings."[37] The sleep disturbances were greatest for those who were on social media or who were surfing the internet in bed.[38]

It's not just social media on smartphones that has disturbed sleep for Gen Z; sleep deprivation is increased by the ease of access to other highly stimulating smartphone activities, including mobile gaming and video streaming.[39] As the CEO of Netflix put it on an earnings call with investors when asked about Netflix's competitors, "You know, think about it, when you watch a show from Netflix and you get addicted to it, you stay up late at night. We're competing with sleep, on the margin."[40]

What does sleep deprivation do to the rapidly changing brains of adolescents? To answer that question, we can turn to the findings of the Adolescent Brain Cognitive Development Study, which scanned the brains of more than 11,000 9- and 10-year-olds back in 2016 and has been following them as they went through puberty and adolescence. Hundreds of academic papers have emerged from this large collaboration, and several examined the effects of sleep deprivation. For example, a 2020 study found that greater sleep disturbance and shorter total sleep time were associated with greater internalizing scores (which include depression), as well as greater externalizing scores (which include aggression and other antisocial actions associated with a lack of impulse control).[41] They also found that the size of the sleep disturbance at the start of the study "significantly predicted depression and internalizing and externalizing scores at 1-year follow-up." In other words, when your sleep is truncated or disturbed, you're more likely to become depressed and develop behavioral problems. The effects were larger for girls.

In short, children and adolescents need a lot of sleep to promote healthy brain development and good attention and mood the next day. When screens are allowed in bedrooms, however, many children will use them late into the night—especially if they have a small screen that can be used under the blanket. The screen-related decline of sleep is likely a contributor to the tidal wave of adolescent mental illness that swept across many countries in the early 2010s.

HARM #3: ATTENTION FRAGMENTATION

Kurt Vonnegut's 1961 short story "Harrison Bergeron" is set in an ultra-egalitarian future America where, by constitutional amendment, nobody is allowed to be smarter, better looking, or more physically able than anyone else. The "handicapper general" is the government officer tasked with enforcing equality of abilities and outcomes. Anyone with a high IQ is required to wear an earpiece at all times that buzzes loudly every 20 seconds or so with a variety of noises designed to interrupt sustained thinking, thereby bringing the person down to the functional intelligence of the average citizen.

I thought about this story as I began to talk with my students a few years ago about how their phones were affecting their productivity. Young people have relied, since the late 1990s, on texting as their basic mode of communication. They keep their ringers off, which means that their phones vibrate repeatedly throughout the day, especially when they participate in group chats. But the situation was far worse than I had imagined. Most of my students get alerts from dozens of apps, including messaging apps (such as WhatsApp), social media apps (Instagram and Twitter), and a variety of news sites that ping them with "breaking news" about politics, sports, and the romantic lives of celebrities. For my MBA students (who are mostly in their late 20s), there are also work-related apps such as Slack. Most of my students also have their phones set to vibrate with an alert every time an email message arrives.

When you add it all up, the average number of notifications on young people's phones from the top social and communication apps amounts to 192 alerts per day, according to one study.[42] The average teen, who now gets only seven hours of sleep per night, therefore gets about 11 notifications per waking hour, or one every five minutes. And that's just for the apps that are about communication. When we add in the dozens of other apps for which they have not turned off push notifications, the number of interruptions grows far higher. And we're still only talking about the *average* teen. If we zoom in on heavy users, such as older teen girls, who use texting and social media apps far more often than any other group, we are now in the ballpark of one interruption every minute. Thanks to the tech industry and its voracious competition for the limited resource of adolescent attention, many members of Gen Z are now living in Kurt Vonnegut's dystopia.

In 1890, the great American psychologist William James described attention as "the taking possession by the mind, in clear and vivid form, of one out of what seem several simultaneously possible objects or trains of thought. . . . It implies withdrawal from some things in order to deal effectively with others."[43] Attention is a choice we make to stay on one task, one line of thinking, one mental road, even as attractive off-ramps beckon. When we fail to make that choice and allow ourselves to be frequently sidetracked, we end up in "the confused, dazed, scatterbrained state" that James said is the opposite of attention.

Staying on one road got much harder when the internet arrived and moved much of our reading online. Every hyperlink is an off-ramp, calling us to abandon the choice we made moments earlier. Nicholas Carr, in his aptly titled 2010 book, *The Shallows: What the Internet Is Doing to Our Brains*, lamented his lost ability to stay on one path. Life on the internet changed how his brain sought out information, even when he was offline trying to read a book. It reduced his ability to focus and reflect because he now craved a constant stream of stimulation: "Once I was a scuba diver in the sea of words. Now I zip along the surface like a guy on a Jet Ski."[44]

Carr's book was about the internet as he experienced it on his computers in the 1990s and 2000s. He occasionally mentions BlackBerrys and iPhones, which had become popular just a few years before his book was published. But a buzzing smartphone is so much more alluring than a passive hyperlink, so much deadlier for concentration. Every app is an off-ramp; every notification is a Las Vegas–style sign calling out to you to turn the wheel: "Tap here and I'll tell you what someone just said about you!"

And no matter how hard it is for an adult to stay committed to one mental road, it is far harder for an adolescent, who has an immature frontal cortex and therefore limited ability to say no to off-ramps. James described children like this: "Sensitiveness to immediately exciting sensorial stimuli characterizes the attention of childhood and youth. . . . the child seem[s] to belong less to himself than to every object which happens to catch his notice." Overcoming this tendency to flit around is "the first thing which the teacher must overcome." This is why it is so important that schools go phone-free for the entire school day by using phone lockers or lockable pouches.[45] Capturing the child's attention with "immediately exciting sensorial stimuli" is the goal of app designers, and they are very good at what they do.

This never-ending stream of interruptions—this constant fragmentation of attention—takes a toll on adolescents' ability to think and may leave permanent marks in their rapidly reconfiguring brains. Many studies find that students with access to their phones use them in class and pay far less attention to their teachers.[46] People can't really multitask; all we can do is shift attention back and forth between tasks while wasting a lot of it on each shift.[47]

But even when students don't check their phones, the mere presence of a phone damages their ability to think. In one study, researchers brought college students into the lab and randomly assigned them to (1) leave their bag and phone out in the entry room of the lab, (2) keep their phone with them in their pocket or bag, or (3) put their phone on their desk next to them. They then had the students complete tasks that tested their fluid intelligence and working memory capacity, such as by solving math problems while also remembering a string of letters. They found that performance was best when phones were left in the other room, and worst when phones were visible, with pocketed phones in between. The effect was bigger for heavy users. The article was titled "Brain Drain: The Mere Presence of One's Own Smartphone Reduces Available Cognitive Capacity."[48]

When adolescents have continuous access to a smartphone at that developmentally sensitive age, it may interfere with their maturing ability to focus. Studies show that adolescents with attention deficit hyperactivity disorder (ADHD) are heavier users of smartphones and video games, and the commonsense assumption is that people with ADHD are more likely to seek out the stimulation of screens and the enhanced focus that can be found in video games. But does causation run in the reverse direction too? Can a phone-based childhood exacerbate existing ADHD symptoms?

It appears so.[49] A Dutch longitudinal study found that young people who engaged in more problematic (addictive) social media use at one measurement time had stronger ADHD symptoms at the next measurement time.[50] Another study by a different group of Dutch researchers used a similar design and also found evidence suggesting that heavy media multitasking caused later attention problems, but they found this causal effect only among younger adolescents (ages 11–13), and it was especially strong for girls.[51]

The brain develops throughout childhood, with an acceleration of change during puberty. One of the main skills that adolescents are expected to develop as they advance through middle school and high school is "executive function," which refers to the child's growing ability to make plans and then do the things necessary to execute those plans. Executive function skills are slow to develop because they are based in large part in the frontal cortex, which is the last part of the brain to rewire during puberty. Skills essential for executive function include self-control, focus, and the ability to resist off-ramps. A phone-based childhood is likely to interfere with the development of executive function.[52] I cannot say that light use of these products is harmful to attention, but among heavy users we do consistently find worse outcomes in part because such users are often, to some degree, addicted.

HARM #4: ADDICTION

When my daughter found herself powerless to lift her eyes up from my iPad, what exactly was going on in her brain? Thorndike didn't know about neurotransmitters, but he correctly guessed that the repetition of small pleasures played a big role in laying down those new paths in the brain. Now we know that when an action is followed by a good outcome (such as gaining food, or relieving pain, or just achieving a goal), certain brain circuits involved with learning release a bit of dopamine—the neurotransmitter most centrally involved in feelings of pleasure and pain. The release of dopamine feels good; we register it in our consciousness.

But it's not a passive reward that satisfies us and reduces our craving. Rather, dopamine circuits are centrally involved in *wanting*, as in "that felt great, I want more!" When you eat a potato chip, you get a small hit of dopamine, which is why you then want the second one even more than you wanted the first one.

It's the same with slot machines: A win feels great, but it doesn't cause gambling addicts to take their earnings and go home, satisfied. Rather, the pleasure motivates them to keep going. It's the same for video games, social media, shopping sites, and other apps that routinely cause people to spend far more time or money than they had intended to spend. The neural basis of behavioral addictions to social media or video games is not exactly the same as chemical addictions to cocaine or opiates.[53] Nonetheless, they all involve dopamine, craving, compulsion, and the feeling my daughter expressed—that she was powerless to act on her conscious wishes. That happens by design. The creators of these apps use every trick in the psychologists' tool kit to hook users as deeply as slot machines hook gamblers.[54]

To be clear, the great majority of adolescents using Instagram or playing *Fortnite* are not addicted, but their desires are being hacked and their actions manipulated nonetheless. Of course, advertisers have long sought to do exactly this, but touch screens and internet connections opened up vast new possibilities for employing behaviorist techniques, which work best when there are rapid cycles or loops of behaviors and rewards. One researcher who explored these possibilities was B.J. Fogg, a professor at Stanford who wrote a 2002 book titled *Persuasive Technology: Using Computers to Change What We Think and Do*. Fogg also taught a course titled "Persuasive Technology" in which he taught students how to take behaviorist techniques for training animals and apply them to humans. Many of his students went on to found or work at social media companies, including Mike Krieger, a cofounder of Instagram.

How do habit-forming products hook adolescents? Take the case of a 12-year-old girl sitting at her desk at home, struggling to understand photosynthesis for a test in her science class the next day. How can Instagram lure her away and then keep her away for an hour? App design-

Figure 5.3. The Hooked model. From Nir Eyal's 2014 book, *Hooked: How to Build Habit-Forming Products*. In the book, Eyal warned about the ethical implications of misusing the model in a section titled "The Morality of Manipulation."[55]

ers often use a four-step process that creates a self-perpetuating loop, shown in Figure 5.3.

The Hooked model guides designers through the loop they need to create if they want to build strong habits in their users.

The loop starts with an external trigger, such as a notification that someone commented on one of her posts. That's step 1, the off-ramp inviting her to leave the path she was on. It appears on her phone and automatically triggers a desire to perform an action (step 2) that had previously been rewarded: touching the notification to bring up the Instagram app. The action then leads to a pleasurable event, but only sometimes, and this is step 3: a variable reward. Maybe she'll find some expression of praise or friendship, maybe not.

This is a key discovery of behaviorist psychology: It's best not to reward a behavior *every time* the animal does what you want. If you reward an animal on a *variable-ratio schedule* (such as one time out of every 10 times, on average, but sometimes fewer, sometimes more), you create the strongest and most persistent behavior. When you put a rat into a cage where it has learned to get food by pressing a bar, it gets a surge of dopamine in anticipation of the reward. It runs to the bar and starts pressing. But if the first few presses yield no reward, that does not dampen the rat's enthusiasm. Rather, as the rat continues to press, dopamine levels will go *up* in anticipation of the reward, which must be coming at any moment! When the reward finally comes, it feels great, but the heightened levels of dopamine make the rat continue to press, in anticipation of the next reward, which will come . . . after some unknown number of presses, so just keep pressing! There is no off-ramp in an app with a bottomless feed; there is no signal to stop.

These first three steps are classic behaviorism. They deploy operant conditioning as taught by B. F. Skinner in the 1940s. What the Hooked model adds for humans, which was not applicable for those working with rats, was the fourth step: investment. Humans can be offered ways to put a bit of themselves into the app so that it matters more to them. The girl has already filled out her profile, posted many photos of herself, and linked herself to all of her friends plus hundreds of other Instagram users. (Her brother, studying for an exam in the room next to hers, has spent hundreds of hours accumulating digital badges, purchased "skins," and made other investments in video games such as *Fortnite* and *Call of Duty.*)

At this point, after investment, the trigger for the next round of behavior may become *internal.* The girl no longer needs a push notification to call her over to Instagram. As she is rereading a difficult passage in her textbook, the thought pops up in her mind: "I wonder if anyone has liked the photo I posted 20 minutes ago?" An attractive off-ramp appears in consciousness (step 1). She tries to resist temptation and stick with her homework, but the mere thought of a possible reward triggers the release of a bit of dopamine, which makes her *want* to go to Instagram immediately. She feels a craving. She goes (step 2) and finds that nobody liked or commented on her post. She feels disappointment, but her dopamine-primed brain still craves a reward, so she starts looking through her other posts, or her direct messages, or anything that shows that she mat-ters to someone else, or anything that provides easy entertainment, which she finds (step 3). She wanders down her feed, leaving comments for her friends along the way. Sure enough, a friend reciprocates by liking her last post. An hour later, she returns to her study of photosynthesis, depleted and less able to focus.

Once the user's own feelings are enough to trigger a behavior that gets variably rewarded, the user is "hooked." We know that Facebook intentionally hooked teens using behaviorist techniques thanks to the Facebook Files—the trove of internal documents and screenshots of presentations brought out by the whistleblower Frances Haugen in 2021. In one chilling section, a trio of Facebook employees give a presentation titled "The Power of Identities: Why Teens and Young Adults Choose Instagram." The stated objective is "to support Facebook Inc.–wide product strategy for engaging younger users." A section titled "Teen Fundamentals" delves into neuroscience, showing the gradual maturation of the brain during puberty, with the frontal cortex not mature until after age 20. A later photo shows an MRI image of a brain with this caption:

> The teenage brain is usually about 80% mature. The remaining 20% rests in the frontal cortex. . . . At this time teens are highly dependent on their temporal lobe where emotions, memory and learning, and the reward system reign supreme.

A subsequent slide shows the loop that Facebook's designers strive to create in users and notes the points of vulnerability (see Figure 5.4).

Many other slides in the presentation indicate that the presenters were not trying to protect the young woman in the center from overuse and addiction; their goal was to advise other Facebook employees on how to keep her "engaged" for longer with rewards, novelty, and emotions. Suggestions include making it easier for teens to open multiple accounts and implementing "stronger paths to related interest content."

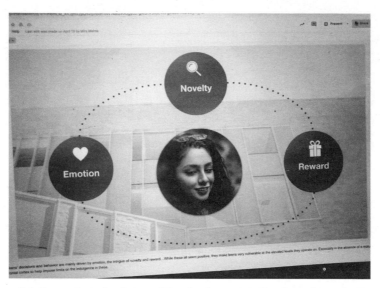

Figure 5.4. Screenshot of an internal Facebook presentation, brought out by Frances Haugen. The caption says, "Teens' decisions and behavior are mainly driven by emotion, the intrigue of novelty and reward. While these all seem positive, they make teens very vulnerable at the elevated levels they operate on. Especially in the absence of a mature frontal cortex to help impose limits on the indulgence in these." (Source: The Facebook Files, section 42/15, p. 53.)[56]

IN HER BOOK *DOPAMINE NATION*, THE STANFORD UNIVERSITY ADDICTION researcher Anna Lembke explains how addiction plays out in her patients, who suffer from a variety of drug and behavioral addictions (such as gambling, shopping, sex). Increasingly in the 2010s, she began to treat teenagers who had digital addictions. Like people with heroin and cocaine addictions, those addicted to digital activities found that "nothing feels good anymore" when they were not doing their preferred activity. The reason is that the brain adapts to long periods of elevated dopamine by changing itself in a variety of ways to maintain homeostasis. The most important adaptation is by "downregulating" dopamine transmission. The user needs to increase the dosage of the drug to get the pleasure back.

Unfortunately, when an addicted person's brain adapts by counteracting the effect of the drug, the brain then enters a state of deficit when the user is *not* taking the drug. If dopamine release is pleasurable, dopa-

mine deficit is unpleasant. Ordinary life becomes boring and even painful without the drug. Nothing feels good anymore, except the drug. The
addicted person is in a state of withdrawal, which will go away only if she
can stay off the drug long enough for her brain to return to its default
state (usually a few weeks).

Lembke says that "the universal symptoms of withdrawal from any
addictive substance are anxiety, irritability, insomnia, and dysphoria."[57]
Dysphoria is the opposite of euphoria; it refers to a generalized feeling of
discomfort or unease. This is basically what many teens say they feel—
and what parents and clinicians observe—when kids who are heavy
users of social media or video games are separated from their phones
and game consoles involuntarily. Symptoms of sadness, anxiety, and irritability are listed as the signs of withdrawal for those diagnosed with
internet gaming disorder.[58]

Lembke's list of the universal symptoms of withdrawal shows us
how addiction magnifies the three other foundational harms. Most obviously, those who are addicted to screen-based activities have more trouble falling asleep, both because of the direct competition with sleep and
because of the high dose of blue light delivered to the retina from just
inches away, which tells the brain: It's morning time! Stop making mela-
tonin![59] Also, while most people wake up multiple times during the night
and then fall right back to sleep, people who have become addicted will
often reach for their phones and start scrolling.

Lembke writes, "The smartphone is the modern-day hypodermic
needle, delivering digital dopamine 24/7 for a wired generation."[60] Her
metaphor helps to explain why the transition from play-based childhood
to phone-based childhood has been so devastating, and why the crisis
showed up so suddenly in the early 2010s. Millennial adolescents in the
1990s and early 2000s had access to all kinds of addictive activities on
their home computers, and some of them did get addicted. But they
couldn't take their computers with them everywhere they went. After
the Great Rewiring, the next generation of adolescents could, and did.

To see the far-reaching effects of the transition to smartphones,
imagine a sleep-deprived, anxious, and irritable student interacting

with fellow students at school. It's not likely to go well, especially if her school allows her to keep her phone with her during the school day. She'll use much of lunchtime and time between classes to catch up on social media, rather than having the synchronous, face-to-face hangout time she needs for healthy social development, thereby further compounding her feelings of social isolation.

Now imagine a sleep-deprived, anxious, irritable, and socially isolated student trying to focus on her homework as off-ramps beckon from the phone lying faceup on her desk. Her impaired executive abilities will strain to keep her on task for more than a minute or two at a time. Her attention is fragmented. Her consciousness becomes "the confused, dazed, scatterbrained state" that William James said is the opposite of attention.

When we gave children and adolescents smartphones in the early 2010s, we gave companies the ability to apply variable-ratio reinforcement schedules all day long, training them like rats during their most sensitive years of brain rewiring. Those companies developed addictive apps that sculpted some very deep pathways in our children's brains.[61]

ON THE BENEFITS OF SOCIAL MEDIA FOR ADOLESCENTS

In 2023, the U.S. surgeon general, Vivek Murthy, issued an advisory discussing the effects of social media use on youth mental health.[62] The advisory warned that social media poses "a profound risk of harm to the mental health and well-being of children and adolescents." His 25-page report outlined the potential costs and benefits of social media use. Regarding benefits, he stated,

> Social media can provide benefits for some youth by providing positive community and connection with others who share identities, abilities, and interests. It can provide access to important information and create a space for self-expression. The ability to form and maintain friendships online and develop social connec-

tions are among the positive effects of social media use for youth. These relationships can afford opportunities to have positive interactions with more diverse peer groups than are available to them offline and can provide important social support to youth. The buffering effects against stress that online social support from peers may provide can be especially important for youth who are often marginalized, including racial, ethnic, and sexual and gender minorities.

These benefits all sound plausible, and indeed the surgeon general was drawing on surveys showing that many teens say that they obtain these benefits from social media. For instance, a 2023 Pew report found that 58% of teenagers report that social media helps them feel more accepted, 71% saw it as a creative outlet, and 80% felt more in touch with their friends' lives.[63] A 2023 Common Sense Media report found that 73% of girls report having fun daily on TikTok, and 34% said their lives would be worse if they did not have access to the platform. Sixty-three percent of girls say they have fun daily on Instagram, with 21% saying their lives would be worse without it.[64]

Certainly, these digital platforms offer fun and entertainment, as television did for previous generations. They also confer some unique benefits for specific groups such as sexual minority youth and those with autism—where some virtual communities can help soften the pain of social exclusion in the real world.[65]

However, unlike the extensive evidence of harm found in correlational, longitudinal, and experimental studies, there is very little evidence showing *benefits* to adolescent mental health from long-term or heavy social media use.[66] There was no wave of mental health and happiness breaking out around the world in 2013, as young people embraced Instagram. Teens are certainly right when they say that social media gives them a connection with their friends, but as we've seen in their reports of increasing loneliness and isolation, that connection does not seem to be as good as what it replaced.

A second reason why I am skeptical of claims about the benefits of social media for adolescents is that these claims often confuse social media with the larger internet. During the COVID shutdowns I often heard people say, "Thank goodness for social media! How would young people have connected without it?" To which I respond: Yes, let's imagine a world in which the only way that children and adolescents could connect was by telephone, text, Skype, Zoom, FaceTime, and email, or by going over to each other's homes and talking or playing outside. And let's imagine a world in which the only way they could find information was by using Google, Bing, Wikipedia, YouTube,[67] and the rest of the internet, including blogs, news sites, and the websites of the many nonprofit organizations devoted to their specific interests.[68]

A third reason for skepticism is that the same demographic groups that are widely said to benefit most from social media are also the *most* likely to have bad experiences on these platforms. The 2023 Common Sense Media survey found that LGBTQ adolescents were *more* likely than their non-LGBTQ peers to believe that their lives would be better without each platform they use.[69] This same report found that LGBTQ girls were more than twice as likely as non-LGBTQ girls to encounter harmful content related to suicide and eating disorders. Regarding race, a 2022 Pew report found that Black teens were about twice as likely as Hispanic or white teens to say they think their race or ethnicity made them a target of online abuse.[70] And teens from low-income households ($30,000 or less) were twice as likely as teens from higher-income families ($75,000 or higher) to report physical threats online (16% versus 8%).

My fourth reason for skepticism is that these discussions of benefits rarely consider the age of the child. All of the benefits sound plausible for older teens, but do we really think that 12-year-olds need Instagram or TikTok to "connect" them with strangers instead of simply seeing their friends in person? I cannot see any justification for not enforcing the current minimum age of 13 for opening accounts on social media platforms.

We need to develop a more nuanced mental map of the digital landscape. Social media is not synonymous with the internet, smartphones are not equivalent to desktop computers or laptops, PacMan is not World of Warcraft, and the 2006 version of Facebook is not the 2024 version of

TikTok. Almost all of it is more harmful to preteens than to older teens. I'm not saying that 11-year-olds should be kept off the internet. I'm saying that the Great Rewiring of Childhood, in which the phone-based childhood replaced the play-based childhood, is the major cause of the international epidemic of adolescent mental illness. We need to be careful about which kids have access to which products, at which ages, and on which devices. Unfettered access to everything, everywhere, at any age has been a disaster, even if there are a few benefits.

IN SUM

In this chapter I described the four foundational harms of the phone-based childhood. These are profound changes to childhood caused by the rapid technological shift of the early 2010s. Each one is foundational because it affects the development of multiple social, emotional, and cognitive abilities.

- The sheer amount of time that adolescents spend with their phones is staggering, even compared with the high levels of screen time they had before the invention of the iPhone. Studies of time use routinely find that the average teen reports spending more than seven hours a day on screen-based leisure activities (not including school and homework).

- The opportunity cost of a phone-based childhood refers to everything that children do less of once they get unlimited round-the-clock access to the internet.

- The first foundational harm is social deprivation. When American adolescents moved onto smartphones, time with friends in face-to-face settings plummeted immediately, from 122 minutes per day in 2012 down to 67 minutes per day in 2019. Time with friends dropped further because of COVID restrictions, but Gen Z was already socially distanced before COVID restrictions were put in place.

- The second fundamental harm is sleep deprivation. As soon as adolescents moved from basic phones to smartphones, their sleep declined in both quantity and quality, around the developed world. Longitudinal studies show that smartphone use came first and was followed by sleep deprivation.

- Sleep deprivation is extremely well studied, and its effects are far reaching. They include depression, anxiety, irritability, cognitive deficits, poor learning, lower grades, more accidents, and more deaths from accidents.

- The third fundamental harm is attention fragmentation. Attention is the ability to stay on one mental road while many off-ramps beckon. Staying on a road, staying on a task, is a feature of maturity and a sign of good executive function. But smartphones are kryptonite for attention. Many adolescents get hundreds of notifications per day, meaning that they rarely have five or 10 minutes to think without an interruption.

- There is evidence that the fragmentation of attention in early adolescence caused by problematic use of social media and video games may interfere with the development of executive function.

- The fourth fundamental harm is addiction. The behaviorists discovered that learning, for animals, is "the wearing smooth of a path in the brain." The developers of the most successful social media apps used advanced behaviorist techniques to "hook" children into becoming heavy users of their products.

- Dopamine release is pleasurable, but it does not trigger a feeling of satisfaction. Rather, it makes you want more of whatever you did to trigger the release. The addiction researcher Anna Lembke says that the universal symptoms of withdrawal are "anxiety, irritability, insomnia, and dysphoria." She and other researchers find that many adolescents have developed behavioral addictions that are very much like the way that people develop addictions to slot machine gambling, with profound consequences for their well-being, their social development, and their families.

- When we put these four foundational harms together, they explain why mental health got so much worse so suddenly as soon as childhood became phone-based.

Chapter 6

WHY SOCIAL MEDIA HARMS
GIRLS MORE THAN BOYS

Alexis Spence was born on Long Island, New York, in 2002. She got her first iPad for Christmas in 2012, when she was 10. Initially she used it for Webkinz—a line of stuffed animals that enables children to play with a virtual version of their animal. But in 2013, while in fifth grade, some kids teased her for playing this childish game and urged her to open an Instagram account.

Her parents were very careful about technology use. They maintained a strict prohibition on screens in bedrooms; Alexis and her brother had to use a shared computer in the living room. They checked Alexis's iPad regularly, to see what apps she had. They said no to Instagram.

Like many young users, however, Alexis found ways to circumvent those rules. She opened an Instagram account herself by stating that she was 13, even though she was 11. She would download the app, use it for a while, and then delete it so her parents wouldn't see it. She learned, from other underage Instagram users, how to hide the app on her home screen under a calculator icon, so she no longer had to delete it. When her parents eventually learned that she had an account and began to monitor it and set restrictions, Alexis made secondary accounts where she could post without their knowledge.

At first, Alexis was elated by Instagram. In November 2013, she wrote in her journal, "On Instagram I reach 127 followers. Ya! Let's put it this way, if I was happy and excited for 10 followers then this is just AMAZING!!!!" But over the next few months her mental health plunged, and she began to show signs of depression. Five months after she opened her first Instagram account, she drew the picture in Figure 6.1.

Within six months of opening her account, the content Instagram's algorithms chose for Alexis had morphed from her initial interest in fitness to a stream of photos of models to dieting advice and then to pro-anorexia content. In eighth grade, she was hospitalized for anorexia and depression. She battled eating disorders and depression for the rest of her teen years.

Figure 6.1. Drawing made by Alexis Spence in April 2015, age 12. The words on her laptop are "worthless, die, ugly, stupid, kill yourself." The words on her phone are "stupid, ugly, fat." Copied from the court filing in *Spence v. Meta*.[1]

Alexis is now 21. She has regained control of her life and works as an emergency medical technician, though she still struggles with eating disorders. I spoke with Alexis and her mother after reading the lawsuit her parents filed against Meta for the dangerous product it offered to their daughter, without their permission. I learned more about the dark years when Alexis was in and out of hospitals and about her parents' struggle to keep her away from social media. During one period of separation from social media, Alexis punched a hole in a wall, out of anger. But after a longer hospital stay with no social media, her mother says, Alexis returned to her old sweet self. "She was a different person. She was kind; she was polite. It happened to be during Mother's Day, and she made me the most beautiful Mother's Day card. We had our daughter back."

Why does social media have such a magnetic pull on girls? How, exactly, does it suck them in and then harm many of them, causing depression, anxiety disorders, eating disorders, and suicidal thoughts?[2]

AS WE'VE SEEN, THE MOVE FROM BASIC PHONES TO SMARTPHONES IN THE early 2010s brought a large increase in the variety and intensity of digital activities, as well as related increases in the four foundational harms: social deprivation, sleep deprivation, attention fragmentation, and addiction. Around 2013, psychiatric wards in the United States and other Anglo countries began to fill disproportionately with girls.[3] In this chapter, I'll explore the reasons why social media has harmed girls more than boys. In the next chapter, on boys, I'll talk about their different technology use, and I'll show that the hit to their well-being is seen less in their rates of mental illness (which did increase) and more in their declining success and increasing disengagement from the real world. In both chapters, I'll focus on data from the United States and the U.K. because it is so plentiful.[4]

THE EVIDENCE THAT SOCIAL MEDIA HARMS GIRLS

Social media platforms, as I defined them in the previous chapter, serve the function of sharing user-generated content widely and asynchronously.

On the most prototypical platforms, like Instagram, users post content—often about themselves—and then wait for the judgments and comments of others. Such posting and waiting, along with social comparison, is having larger and more harmful effects on girls and young women than on boys and young men, and this difference shows up consistently in many correlational studies. These studies typically ask teens about their technology use, and also about their mental health, and then look to see whether those who use more of a certain technology end up with worse mental health.

I should note that some studies have failed to find evidence of harm. One well-known study reported that the association of digital media use with harmful psychological outcomes was so close to zero that it was roughly the same size as the association of "eating potatoes" with such harms.[5] But when Jean Twenge and I reanalyzed the same data sets and zoomed in on the association of *social media* (as opposed to a broader measure of digital technology use that included watching TV and owning a computer) with poor mental health for *girls* (instead of all teens merged together), we found much larger correlations.[6] The proper comparison was no longer eating potatoes but instead binge drinking or using marijuana. There is a clear, consistent, and sizable link[7] between heavy social media use and mental illness for girls,[8] but that relationship gets buried or minimized in studies and literature reviews that look at all digital activities for all teens.[9] Journalists who report that the evidence of harm is weak are usually referring to such studies.[10]

You can see this sizable link in figure 6.2, which reports data from the Millennium Cohort Study, a British study that followed roughly 19,000 children born around the year 2000 as they matured through adolescence. The figure shows the percentage of U.K. teens who could be considered depressed (based on their responses to a 13-item depression scale), as a function of how many hours they reported spending on social media on a typical weekday. For boys, knowing how much time they spend on social media doesn't tell you much, unless they say that they are heavy users. It's only when boys are spending more than two hours a day that the curve begins to rise.

Depression by Level of Social Media Use, U.K.

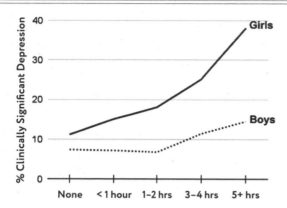

Figure 6.2. Percent of U.K. teens depressed as a function of hours per weekday on social media. Teens who are heavy users of social media are more depressed than light users and nonusers, and this is especially true for girls. (Source: Millennium Cohort Study.)[11]

For girls, there is a larger and more consistent relationship. The more time a girl spends on social media, the more likely she is to be depressed. Girls who say that they spend five or more hours each weekday on social media are *three times* as likely to be depressed as those who report no social media time.

Is Social Media a Cause or Just a Correlate?

Correlational studies are always open to multiple interpretations. There could be "reverse correlation," in which depression drives girls to use social media, rather than the other way around.[12] There could be a third variable, such as genetics, or overly permissive parenting, or loneliness, that causes both. To establish that one thing *caused* another to happen, the main tool scientists use is an experiment in which some people are randomly assigned to receive a treatment and other people are randomly assigned to be in the control group, which receives a placebo (in medical studies) or carries on with business as usual (in many social science experiments). Experiments like this are sometimes referred to as RCTs (randomized controlled trials). In some social media experiments, the treatment is to require young adults to go for a few days or weeks with reduced or no social media access. In other experiments, the treatment is to bring young adults (usually college students) into the lab and put them in situations that model some aspect of being on social media (such as swiping through photos) and then see how these treatments affect psychological variables.

For example, one study randomly assigned college students to greatly reduce the use of social media platforms (or not reduce, for the control group) and then measured their depressive symptoms three weeks later. The authors reported that "the limited use group showed significant reductions in loneliness and depression over three weeks compared to the control group."[13] Another study randomly assigned teen girls to be exposed to selfies taken from Instagram, either in their original state or after modification by the experimenters to be extra attractive. "Results showed that exposure to manipulated Instagram photos directly led to lower body image."[14] Taken as a whole, the dozens of experiments that Jean Twenge, Zach Rausch, and I have collected[15] confirm and extend the patterns found in the correlational studies: Social media use is a *cause* of anxiety, depression, and other ailments, not just a *correlate*.

Does Social Media Affect Groups as Well as Individuals?

There is a major limitation to all of these experiments: They look for effects of social media on *individuals in isolation*, as if we were studying the health effects of consuming sugar. If 100 teens are randomly assigned to reduce their sugar intake for three months, will they experience any health benefits, compared with a control group? But social media is not like sugar. It doesn't just affect the person who consumes it. When it was carried into schools in the early 2010s, on smartphones in students' pockets, it quickly changed the culture for everyone. (Communication networks rapidly become more powerful as they grow.[16]) Students talked to each other less between classes, at recess, and at lunch, because they began to spend much of that time checking their phones, often getting caught up in microdramas throughout the day.[17] This meant that they made eye contact less frequently, laughed together less, and lost practice making conversation. Social media therefore harmed the social lives even of students who stayed away from it.

These group-level effects may be much larger than the individual-level effects, and they are likely to suppress the true size of the individual-level effects.[18] If an experimenter assigns some adolescents to abstain from social media for a month while all of their friends are still on it, then the abstainers are going to be more socially isolated for that month. Yet even still, in several studies, getting off social media improves their mental health. So just imagine how much bigger the effect would be if *all* of the students in 20 middle schools could be randomly assigned to give up social media for a year, or (more realistically) to put their phones in a phone locker each morning, while 20 other middle schools served as the control group. These are the kinds of experiments we most need in order to examine group-level effects.

There is one small but important class of experiments that *does* measure group-level effects by asking, how does a *whole community* change when social media suddenly becomes much more available in that community?[19] For example, one study took advantage of the fact that Facebook was originally offered only to students at a small number of colleges. As the company expanded to new colleges, did mental health change in the following year or two at those institutions, compared with colleges where students did not yet have access to Facebook? Yes, it got worse, with bigger effects on women. The authors found that

> the roll-out of Facebook at a college increased symptoms of poor mental health, especially depression, and led to increased utilization of mental healthcare services. We also find that, according to the students' reports, the decline in mental health translated into worse academic performance. Additional evidence on mechanisms suggests the results are due to Facebook fostering unfavorable social comparisons.[20]

I have found five studies that looked at the rollout of high-speed internet around the world, and *all five* found evidence of damage to mental health. It's hard to have a phone-based childhood when data speeds are low. For example, what happened in Spain as fiber-optic cables were laid and high-speed internet came to different regions at different times? A 2022 study analyzed "the effect of access to high-speed Internet on hospital discharge diagnoses of behavioral and mental health cases among adolescents." The conclusion:

> We find a positive and significant impact on girls but not on boys. Exploring the mechanism behind these effects, we show that [the arrival of high-speed internet] increases addictive Internet use and significantly decreases time spent sleeping, doing homework, and socializing with family and friends. Girls again power all these effects.[21]

These studies, and many more,[22] indicate that the rapid movement of adolescent social life onto social media platforms was a *cause*, not just a correlate, of the increase in depression, anxiety, suicidal thoughts, and other mental health problems that began in the early 2010s.[23] When some researchers say that the correlations or effect sizes are too small to explain such large increases, they are referring to studies that measured *only individual-level effects*. They rarely consider the rapid transformation of group dynamics that I am calling the Great Rewiring of Childhood.

But why does social media hurt girls more than boys, specifically? How does it affect their developing brains and identities?

GIRLS USE SOCIAL MEDIA MORE THAN BOYS

In the early 2010s, thanks to smartphones, boys and girls started spending more time online, but they spent their time differently. Boys gravitated to watching YouTube videos, to using text-based platforms such as Reddit, and especially to playing online multiplayer video games. Girls became much heavier users of the new visually oriented platforms, primarily Instagram, followed by Snapchat, Pinterest, and Tumblr.[24]

A 2017 study in the U.K. asked teens to rate the effects of the most popular social media platforms on different parts of their well-being, including anxiety, loneliness, body image, and sleep. Teens rated Instagram as the worst of the five apps, followed by Snapchat. YouTube was the only platform that received a positive overall score.[25]

The visually oriented platforms all used the business model developed by Facebook: Maximize time spent on the platform in order to maximize the extraction of data and the value of the user to advertisers. Figure 6.3 shows the percentage of American high school students who spent more than *40 hours a week* using social media platforms. That's like working a full-time job while also being a full-time student. By 2015, one in seven American girls had reached this astronomical level. The survey question was only added in 2013. If we had the data going back to 2010, when few teens had smartphones, the numbers would likely have been close to zero. It was almost impossible for teens to spend 40 hours a week on social media before they could carry the internet with them in their pockets.[26]

Girls spend more time on social media platforms,[27] and the platforms

Social Media Super-Users (40+ Hours per Week)

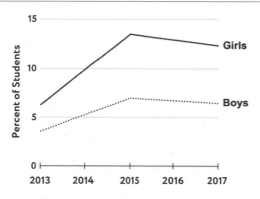

Figure 6.3. Percent of U.S. students (8th, 10th, and 12th grade) who reported spending 40 or more hours a week on social media. (Source: Monitoring the Future.)[28]

they are on are the worst for mental health. So even if girls and boys were identical psychologically, we would expect to find larger increases in anxiety and depression among girls.

But girls and boys are not identical psychologically. There are a number of reasons why girls' core developmental needs are more easily exploited and subverted by social media than is the case for boys (whose needs are more easily exploited by video game companies).

AGENCY AND COMMUNION

Girls and boys are similar psychologically in most ways; an introductory psychology textbook needs only occasional notes about gender differences. But there are a few gender differences that show up widely across cultures and eras. One that is useful for understanding media effects is the distinction between agency and communion, which refer to two sets of motivations or goals found in nearly everyone. One recent review defined them in this way:

> Agency arises from striving to individuate and expand the self and involves qualities such as efficiency, competence, and assertiveness. Communion arises from striving to integrate the self in a larger social unit through caring for others and involves qualities such as benevolence, cooperativeness, and empathy.[29]

The two motives are woven together in changing patterns across the life course, and that weaving is particularly important for adolescents who are developing their identities. Part of defining the self comes from successfully integrating into groups; part of being attractive to groups is demonstrating one's value as an individual with unique skills.[30]

Researchers have long found that boys and men are more focused on agency strivings while girls and women are more focused on communion strivings.[31] The fact that these gender differences have decreased over time tells us that they result in part from cultural factors and forces. The fact that they emerge early in children's play[32] and can be found in the gendered play patterns of other primates[33] tells us that there is probably a biological contribution as well. For our purposes in this book it doesn't matter where the difference comes from. What matters is that tech companies know about these differences and use them to hook their core audience. Social media offers new and easy ways to "connect." Social media *seems* to satisfy communion needs, but in many ways it frustrates them.

FOUR REASONS WHY GIRLS ARE PARTICULARLY VULNERABLE

There are at least four ways that social media companies exploit girls' greater need for communion and their other social concerns. I believe these pathways, viewed together, explain why girls' mental health collapsed so quickly, in so many countries, as soon as they got smartphones and moved their social lives onto Instagram, Snapchat, Tumblr, and other "sharing" platforms.

Reason #1: Girls Are More Affected by Visual Social Comparison and Perfectionism

The 2021 song "Jealousy, Jealousy" by Olivia Rodrigo sums up what it's like for many girls to scroll through social media today. The song begins,

> *I kinda wanna throw my phone across the room*
> *'Cause all I see are girls too good to be true.*

Rodrigo then says that "co-comparison" with the perfect bodies and paper-white teeth of girls she doesn't know is slowly killing her. It's a powerful song; I hope you'll listen to it.[34]

Psychologists have long studied social comparison and its pervasive effects. The social psychologist Susan Fiske says that humans are "comparison machines."[35] Mark Leary, another social psychologist, describes the machinery in more detail: It's as if we all have a "sociometer" in our brains—a gauge that runs from 0 to 100, telling us where we stand in the local prestige rankings, moment by moment. When the needle drops, it triggers an alarm—anxiety—that motivates us to change our behavior and get the needle back up.[36]

Teens are particularly vulnerable to insecurity because their bodies and their social lives are changing so rapidly as they leave childhood. They struggle to figure out where they fit in the new prestige order for their sex. Nearly all adolescents care how they look, especially as they begin to develop romantic interests. All know that they will be chosen or passed over based in part on their appearance. But for adolescent girls, the stakes are higher because a girl's social standing is usually more closely tied to her beauty and sex appeal than is the case for boys. Compared with boys, when girls go onto social media, they are subjected to more severe and constant judgments about their looks and their bodies, and they're confronted with beauty standards that are further out of reach.

It was bad enough when I was growing up in the 1970s and 1980s, when girls were exposed to airbrushed and later photoshopped models. But those were adult strangers; they were not a girl's competition. So what happened when most girls in a school got Instagram and Snapchat accounts and started posting carefully edited highlight reels of their lives and using filters and editing apps to improve their virtual beauty and online brand? Many girls' sociometers plunged, because most were now below what appeared to them to be the average. All around the developed world, an anxiety alarm went off in girls' minds, at approximately the same time.

You can see the power of filters and tuning apps in Figure 6.4, in which Instagram influencer Josephine Livin demonstrates how easy it is to essentially turn a dial and morph oneself into an ever more unrealistic Instagram beauty.

These tuning apps gave girls the ability to present themselves with perfect skin, fuller lips, bigger eyes, and a narrower waist (in addition to showcasing the most "perfect" parts of their lives).[38] Snapchat offered similar features through its filters, first released in 2015, many of which gave users full lips, petite noses, and doe eyes at the touch of a button.

You can see sociometers plunging in figure 6.5, which shows the per-

Figure 6.4. Beauty filters can make you as perfect as you want to be, which then increases the pressure on other girls to improve their beauty. (Source: Josephine Livin, @josephinelivin, on Instagram.)[37]

centage of American high school seniors who said they were satisfied or completely satisfied when asked the question "how satisfied are you with yourself?" The plunge happened for boys too, and I'll discuss how their lives changed in the next chapter.

Satisfied with Oneself

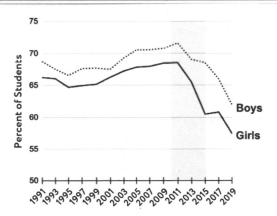

Figure 6.5. The sociometer plunge of 2012. Percent of U.S. students (8th, 10th, and 12th grade) who said they were satisfied with themselves. (Source: Monitoring the Future.)

Girls are especially vulnerable to harm from constant social comparison because they suffer from higher rates of one kind of perfectionism: *socially prescribed* perfectionism, where a person feels that they must live up to very high expectations prescribed by others, or by society at large.[39] (There's no gender difference on *self-oriented perfectionism*, where you torture yourself for failure to live up to your *own* very high standards.) Socially prescribed perfectionism is closely related to anxiety; people who suffer from anxiety are more prone to it. Being a perfectionist also increases your anxiety because you fear the shame of public failure from everything you do. And, as you'd expect by this point in the story, socially prescribed perfectionism began rising, across the Anglosphere nations, in the early 2010s.

Jessica Torres, who ran a plus-size fashion blog, wrote an essay titled "How Being a Social Media Influencer Has Impacted My Mental Health." In it she said,

> The hundreds of dollars and time spent on one Instagram photo felt like a waste. Nothing was perfect enough to post. Even though I was preaching self-love, I was doing the complete opposite with myself. I kept comparing my Instagram page to other influencers whose images were prettier. I began to measure my personal worth and the value of my work with the amount of likes my images were getting.[40]

Since the dawn of advertising, young women have been enticed to pursue seemingly "better" versions of themselves. But social media exposes girls to hundreds or even thousands of such images every day, many of which feature girls too good to be true, with perfect bodies living perfect lives. Exposure to so many images is sure to have a negative effect on comparison machines.

Researchers in France exposed young women either to media photographs of very thin women or to media photographs of average-sized women.[41] They found that the young women exposed to images of very thin women became more anxious about their own body and appearance. But here's the surprising thing: The images were flashed on a screen for just 20 milliseconds, too fast for the women to become consciously aware of what they had seen. The authors conclude that "social comparison takes place outside awareness and affects explicit self-evaluations." This means that the frequent reminders girls give each other that social media is not reality are likely to have only a limited effect, because the part of the brain that is doing the comparisons is not governed by the part of the brain that knows, consciously, that they are seeing only edited highlight reels.

A 13-year-old girl on Reddit explained how seeing other girls on social media made her feel, using the same words as Alexis Spence and Olivia Rodrigo:

> i cant stop comparing myself. it came to a point where i wanna kill myself cause u dont want to look like this and no matter what i try im still ugly/feel ugly. i constantly cry about this. it probably started when i was 10, im now 13. back when i was 10 i found a girl on tiktok and basically became obsessed with her. she was literally perfect and i remember being unimaginably envious of her. throughout my pre-teen years, i became "obsessed" with other pretty girls.[42]

Striving to excel can be healthy when it motivates girls to master skills that will be useful in later life. But social media algorithms home in on (and amplify) girls' desires to be beautiful in socially prescribed ways, which include being thin. Instagram and TikTok send them images of very thin women if they show any interest in weight loss, or beauty, or even just healthy eating. Researchers for the Center for Countering Digital Hate created a dozen fake accounts on TikTok, registered to 13-year-old girls, and found that TikTok's algorithm served them tens of thousands of weight-loss videos within a few weeks of joining the platform.[43] The videos included many emaciated young women urging their followers to try extreme diets such as the "corpse bride" diet or the water-only diet. This is what happened to Alexis Spence on Instagram back in 2012.

Facebook itself commissioned a study on how Instagram was affecting teens in the United States and the U.K. The findings were never released, but whistleblower Frances Haugen smuggled out screenshots of internal documents and shared them with reporters at *The Wall Street Journal.* The researchers found that Instagram is particularly bad for girls: "Teens blame Instagram for increases in the rate of anxiety and depression. . . . This reaction was unprompted and consistent across all groups."[44] The researchers also noted that "social comparison is worse" on Instagram than on rival apps. Snapchat's filters "keep the focus on the face," whereas Instagram "focuses heavily on the body and lifestyle."

Reason #2: Girls' Aggression Is More Relational

Boys were long thought to be more aggressive than girls, and if we look only at violence and physical threats, they are.[45] Boys are also more interested in watching stories and movies about sports, fighting, war, and violence, all of which appeal to agency interests and motivations. Traditionally, boys have negotiated who is high and who is low in social status based in part on who could dominate whom if it came to a fight, or who can hurl an insult at whom without fear of violent reprisal. But because girls have stronger communion motives, the way to really hurt another girl is to hit her in her relationships. You spread gossip, turn her friends against her, and lower her value as a friend to other girls. Researchers have found that when you look at "indirect aggression" (which includes damaging other people's relationships or reputations), girls are higher than boys—but only in late childhood and adolescence.[46] A girl who feels her value sinking is a girl experiencing rising anxiety. If her sociometer drop is sharp enough, she may become depressed and consider suicide. For depressed or ostracized teens, physical death offers the end of pain, whereas social death is a living hell.

Studies confirm that as adolescents moved their social lives online, the nature of bullying began to change. One systematic review of studies from 1998 to 2017 found a *decrease* in face-to-face bullying among boys but an *increase* among girls, especially among younger adolescent girls.[47] A separate study of roughly 16,000 Massachusetts high school students from 2006 through 2012 observed no increase in face-to-face bullying for girls and a decrease for boys. However, cyberbullying among girls surged.[48] According to one major U.S. survey, these high rates of cyberbullying have persisted (though have not increased) between 2011 and 2019. Throughout the period, approximately one in 10 high school boys and one in five high school girls experienced cyberbullying each year.[49] In other words, the move online made bullying and harassment a larger part of daily life for girls.

While the *percentage* of teens who reported being cyberbullied each year might not have risen during the 2010s, the *ways* in which students could perpetrate and experience relational aggression has changed as teens joined new platforms that offered new ways to spread rumors and mount attacks. Social media makes it easy for anyone, of any age, to set up multiple anonymous profiles, which can be used for trolling and reputational destruction. All of it takes place in a virtual world that parents and teachers can rarely access or understand. Additionally, as smartphones accompany adolescents to school, to the bathroom, and to bed, so too can their bullies.

In a 2018 *Atlantic* story about bullying on Instagram,[50] Taylor Lorenz tells the story of Mary, a 13-year-old girl who made it onto the cheer team while a friend did not. The (former) friend used many of Instagram's features to damage Mary's standing with other students. "There was literally a group chat on Instagram named Everyone in the Class but Mary," she said. "All they did on there was talk bad about me." The episode caused Mary to have the first panic attacks of her life.

Social media has magnified the reach and effect of relational bullying, placing immense pressure on girls to monitor their words and actions. They are aware that any misstep can swiftly go viral and leave a permanent mark. Social media fuels the insecurity of adolescence, already a period where there is immense concern about the possibility of ostracism, and has thus turned a generation of girls away from discover mode and toward defend mode.

Freya India, a Gen Z British woman who writes about girls and mental health, wrote an essay titled "Social Media's Not Just Making Girls Depressed, It's Making Us Bitchy Too." In the essay she wrote,

> From anonymous Instagram hate pages to full-blown teenage cancel culture campaigns, today's girls can drag each other down in all kinds of creative ways. Then there's passive aggression, today personified by the subtweet, the "soft block" (where you block and then unblock someone in quick succession, so they no longer follow you), the read receipt; even a public tag in an unflattering photo.[51]

Once again, the transition to phone-based life in the early 2010s upended girls' relationships and lives. Puberty was already a fraught time of transition, with heightened need for a few close friends. Then social media came along to make the transition harder by making relational aggression so much easier and status competition so much more pervasive and public. Many suicides by adolescent girls have been directly linked to bullying and shaming facilitated by social media platforms, including platforms such as Ask.fm and NGL (Not Gonna Lie), which were designed explicitly to encourage users to anonymously broadcast their thoughts about other people.[52]

Reason #3: Girls More Easily Share Emotions and Disorders

We all know that our close friends can affect our moods. But did you know that your friends' friends affect you too? The sociologist Nicholas Christakis and the political scientist James Fowler analyzed data from a long-running survey of the residents in Framingham, Massachusetts, called the Framingham Heart Study.[53] The study focused on physical health, but Christakis and Fowler were able to use items in the survey to study the way emotions moved through the community over time. They found that happiness tends to occur in clusters. This was not just because happy people seek each other out. Rather, when one person became happier, it increased the odds that their existing friends would become happier too. Amazingly, it also had an influence on their friends' friends, and sometimes even on their friends' friends' friends. Happiness is contagious; it spreads through social networks.

In a follow-up study, Christakis and Fowler teamed up with the psychiatrist James Rosenquist to see whether negative emotional states, such as depression, also spread in networks, using the same data set.[54] There were two interesting twists in their findings. The first was that bad was stronger than good, as is almost always the case in psychology.[55] Depression was significantly more contagious than happiness or good mental health. The second twist was that depression spread only from women. When a woman became depressed, it increased the odds of depression in her close friends (male and female) by 142%. When a man became depressed, it had no measurable effect on his friends. The authors surmise that the difference is due to the fact that women are more emotionally expressive and more effective at communicating mood states within friendship pairs. When men get together, in contrast, they are more likely to *do* things together rather than talk about what they are feeling.

The Framingham Heart Study ended in 2001, just before the arrival of social media. What do you suppose happened to communities of teens after 2010, when they became far more tightly interconnected than the adults in the Framingham study? Given that depression and anxiety are more contagious than good mental health, and given that girls are more likely to talk about their feelings than boys, we might expect a sudden burst of depression and anxiety as soon as large numbers of girls joined Instagram and other "sharing" platforms, around 2012.

That is exactly what happened, as I showed in chapter 1. In multiple countries, girls' depression rates rose rapidly in the early 2010s. So did their rates of self-harm and psychiatric hospitalization. But depression is not the only disorder spread by social media.

In 1997, Leslie Boss, then a researcher at the U.S. Centers for Disease

Control, published a review of the historical and medical literature on "sociogenic" epidemics.[56] ("Sociogenic" means "generated by social forces," as opposed to biological causation.) Boss noted that there are two variants that recur throughout history. There is an "anxiety variant" in which abdominal pain, headache, dizziness, fainting, nausea, and hyperventilation are the most common symptoms, and there is a "motor variant" in which the most common symptoms are "hysterical dancing, convulsions, laughing, and pseudoseizures." These "dancing plagues," as they are called by historians, occasionally swept through medieval European villages, leading some townspeople to dance until they died from exhaustion.[57] For both variants, when they have occurred in recent decades, medical authorities have been unable to find any kind of toxin or environmental pollution that could have caused these symptoms. What they do find, repeatedly, is that adolescent girls are at higher risk of succumbing to these illnesses than any other groups and that the outbreaks are more likely when there has been some unusual recent stressor or threat to the community.[58]

Boss noted that these epidemics spread along social networks via face-to-face communication. In more recent times, she noted, they spread by reports in the mass media, such as television. Writing in 1997, in the first years of the internet, she offered this prediction: "Development of new approaches in mass communication, most recently the Internet, increase the ability to enhance outbreaks through communication."

She was prescient. When adolescents moved onto image-based social media platforms, especially video-oriented platforms such as YouTube and later TikTok, they hooked themselves up in a way that facilitated the transmission of psychogenic illnesses. As soon as they did this, rates of anxiety and depression surged around the world, particularly among adolescent girls. Much of the adolescent mental illness epidemic may be the direct result of the anxiety variant spreading by two distinct psychological processes. First, there is simple emotional contagion, as described by Fowler and Christakis. People pick up emotions from others, and emotional contagion is especially strong among girls. Second, there is "prestige bias," which is the social learning rule I described in chapter 2:

Don't just copy anyone; first find out who the most prestigious people are, then copy them. But on social media, the way to gain followers and likes is to be more extreme, so those who present with more extreme symptoms are likely to rise fastest, making them the models that everyone else locks onto for social learning. This process is sometimes known as *audience capture*—a process in which people get trained by their audiences to become more extreme versions of whatever it is the audience wants to see.[59] And if one finds oneself in a network in which most others have adopted some behavior, then the other social learning process kicks in too: conformity bias.

When the COVID pandemic arrived in 2020, both the disease and the lockdowns made sociogenic illness more likely. COVID was a global threat and stressor. The lockdowns led teens to spend even more time on social media, especially on TikTok, which was relatively new. TikTok was especially enticing to adolescent girls, and what it encouraged them to do, in its early days, was to practice highly stylized dance moves that they copied from other girls—dances that spread around the world. But TikTok did not just encourage girls to dance. Its advanced algorithm picked up any sign of interest in anything and sent users more of that, often in a more extreme form. Anyone who revealed an interest in mental health was soon inundated with videos of other teens displaying mental illness and receiving social support for doing so.[60] In August 2023, videos with the hashtag #mentalhealth had more than 100 billion views. #Trauma had more than 25 billion.

A group of German psychiatrists led by Kirsten Müller-Vahl[61] noted a sudden increase in young people showing up at clinics claiming to have Tourette's syndrome—a motor disorder in which patients emit pronounced tics, such as heavy blinking or head and neck rotations, and in which they often emit words or sounds involuntarily. The disease is thought to be related to irregularities in a part of the brain called the basal ganglia, which is heavily involved in physical movement. It usually emerges between the ages of 5 and 10, and 80% of those who have it are boys.

But the German psychiatrists could see that almost none of these patients really had Tourette's. The tics were different, there had been no sign of the disease in them when they were young, and, most revealingly, their tics were astonishingly similar. In fact, these patients—mostly young men in this first wave—were mimicking a single German influencer who actually had Tourette's and who demonstrated his tics in his very popular YouTube videos. These included shouting out, "Flying sharks!" and "Heil Hitler!"[62]

The German researchers wrote, "We report the first outbreak of a new type of mass sociogenic illness that in contrast to all previously reported episodes is spread solely via social media. Accordingly, we suggest the more specific term 'mass social media–induced illness.'"

Even though Tourette's is mostly a male disease, once it became a popular disorder on social media, it spread faster among girls. For example, some girls in Anglo countries suddenly developed tic disorders with head shakes and the common tendency to randomly shout the word "beans." This was triggered by one British influencer, Evie, who modeled those behaviors and shouted out the word "beans."[63] One of the main treatments doctors prescribe for the disorder is getting off social media.

There is evidence that several other disorders are spreading sociogenically, especially via sites that feature video posts such as TikTok, YouTube, and Instagram. Dissociative identity disorder (DID) is a condition that used to be known as multiple personality disorder. It was dramatized in the 1957 movie *The Three Faces of Eve*. The person reports that they have within themselves different identities, known as *alters*, which may have very different personalities, moral profiles, genders, sexualities, and ages. There is often a "bad" alter who encourages the person to do bad things to others or to themselves.

DID used to be rare,[64] but since the arrival of TikTok, there has been an increase, primarily among adolescent girls.[65] Influencers portraying multiple personalities have attracted millions of followers, contributing to an escalating trend of self-identifying with the disorder. Asher, a TikTok influencer who describes themself as one of a "system" of 29 identities, has amassed more than 1.1 million followers. The growing interest in DID is further evidenced by the billions of views garnered by hashtags such as #did (2.8 billion), #dissociativeidentitydisorder (1.6 billion), and #didsystem (1.1 billion).[66] Naomi Torres-Mackie, the head of research at the Mental Health Coalition, encapsulated the trend this way: "All of a sudden, all of my adolescent patients think that they have [DID]. . . . And they don't."[67]

The recent growth in diagnoses of gender dysphoria may also be related in part to social media trends. Gender dysphoria refers to the psychological distress a person experiences when their gender identity doesn't align with their biological sex. People with such mismatches have long existed in societies around the world. According to the most recent diagnostic manual of psychiatry,[68] estimates of the prevalence of gender dysphoria in American society used to indicate rates below one in a thousand, with rates for natal males (meaning those who were biological males at birth) being several times higher than for natal females. But those estimates were based on the numbers of people who sought gender reassignment surgery as adults, which was surely a vast underestimate of the underlying population. Within the past decade, the number of individuals who are being referred to clinics for gender dysphoria has been growing rapidly, especially among natal females in Gen Z.[69] In fact, among Gen Z teens, the sex ratio has reversed, with natal females now showing higher rates than natal males.[70]

Some portion of this increase surely reflects the "coming-out" of young people who were trans but either didn't recognize it or were afraid of the social stigma that would attend the expression of their gender identity. Increasing freedom of gender expression and growing awareness of human variation are both forms of social progress. But the fact that gender dysphoria now often appears in social clusters (such as a group of close friends),[71] the fact that parents and those who transition back to their natal sex identify social media as a major source of information and encouragement,[72] and the fact that gender dysphoria is now being diagnosed among many adolescents who showed no signs of it as children[73] all indicate that social influence and sociogenic transmission may be at work as well.

Reason #4: Girls Are More Subject to Predation and Harassment

You know those stories about middle-aged women who befriend adolescent boys on gaming platforms and then send them money and ask for pictures of their penises as a prelude to meeting for sex? Neither do I. Women's sexuality, in its many variations, is rarely predatory in that way.

According to the evolutionary psychologist David Buss,[74] male and female minds come equipped with certain emotional responses and perceptual sensitivities that helped them to "win" at the mating game as it was played long ago. Both sets of cognitive adaptations influence mating and dating today, but the male set makes men more likely to use coercion, trickery, and violence to get sex, and to focus on adolescents as their targets.[75]

In many regions of the virtual world, some men prey upon teen and preteen girls. Older men prey upon boys too, using online dating apps for gay and bisexual men to find them.[76] But brushes with sexual predators are a larger part of internet life for girls than they are for boys, requiring greater reliance on defend mode.[77]

The journalist Nancy Jo Sales followed a number of teenage girls who attended suburban American high schools. In her 2016 book, *American Girls*, she noted that attention from adult men online is frequent, because the apps make little or no effort to restrict interactions between adults and minors. Lily, a high schooler from Garden City, New York, put it this way:

> It's so easy for older predators to go online and just find a girl . . . because girls want the most friends and they want the most followers and likes, so if someone tries to friend them they'll just friend them back right away without even knowing who they are. So even if it's a serial killer, they still friend them back and maybe even start talking to them. It's scary. Especially since a lot of girls will post pictures of themselves like in their bras and bathing suits, and the people they friend back can see those pictures.[78]

Lily and her peers are often exposed to this kind of attention from adult male strangers. But the predation and exploitation also come from their male classmates. Sales described how nude photographs function as a kind of currency in many middle and high schools. A middle school student in New Jersey told her that boys in her grade try to persuade girls to send them nude photos of themselves, which the boys then sell to high school boys in exchange for alcohol. A group of high school girls she spoke to in Florida told her that asking for and sending nudes was commonplace:

> What percentage of girls were sharing nudes? I asked. "Twenty . . . thirty?" they guessed. "The thing is, with boys," Cassy said, "if you don't send them nudes, they say you're a prude." "Or scared," said Maggie. Had a boy ever asked them for nudes? I asked. "Yes," they said. "They blackmail you," Cassy said. "They say, Oh, I have embarrassing pictures of you, if you don't send nudes I'll send them all out on social media."[79]

When girls' nudes are sent around, it can be devastating and often begins a round of cyberbullying. Boys, however, are less likely to suffer when pictures of their penises are shared. In fact, boys often send such photos to girls as bait, to elicit a reciprocal nude photo. A high school girl named Nina tells Sales, "A girl who sends naked pictures, she's a slut, but if a boy does it, everyone just laughs."[80]

Girls on social media platforms such as Instagram and Snapchat are exposed to the direct messages of adult men who seek them out, and also to school cultures in which photos of their naked bodies become a currency for social prestige among boys, a currency that girls pay for with shame. Sexual predation and rampant sexualization mean that girls and young women must be warier, online, than most boys and young men. They are forced to spend more of their virtual lives in defend mode, which may be part of the reason that their anxiety levels went up more sharply in the early 2010s.

QUANTITY OVER QUALITY

Social media, as it is commonly used by teens today, increases the *quantity* of social connections and thereby reduces their quality and their protective nature. Freddie deBoer, an American author and blogger who writes about education, explains why:

> If we're dividing the hours of the day and our mindshare between more and more relationships relative to the past, we're almost certainly investing less in each individual relationship. Digital substitutions for real-world social engagement reduce the drive to be social but don't satisfy emotional needs. . . . I think this created a really powerful trap: this form of interaction superficially satisfied the drive to connect with other people, but that connection was shallow, immaterial, unsatisfying. The human impulse to see other people was dulled without accessing the reinvigorating power of actual human connection.[81]

When everything moved onto smartphones in the early 2010s, both girls and boys experienced a gigantic increase in the *number* of their social ties and in the *time* required to service these ties (such as reading and commenting on the posts of acquaintances or maintaining dozens of Snapchat "streaks" with people who are not your closest friends). This explosive growth necessarily caused a decline in the number and depth of close friendships, which you can see in figure 6.6.

The clinical psychologist Lisa Damour says that regarding friendship for girls, "quality trumps quantity." The happiest girls "aren't the ones who have the most friendships but the ones who have strong, supportive friendships, even if that means having a single terrific friend."[82] (She notes that this is true for boys as well.) Once girls flocked onto social media platforms and had fewer long talks with one or two special friends, they found themselves immersed in a vast sea of transient, unreliable, fair-weather "friends," followers, and acquaintances. Quantity trumped

Have a Few Close Friends

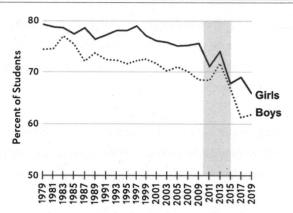

Figure 6.6. The percentage of U.S. high school seniors who agreed or mostly agreed with the statement "I usually have a few friends around that I can get together with." Rates dropped slowly before 2012, and more quickly afterward. (Source: Monitoring the Future.)[83]

quality and loneliness surged, as you can see in figure 6.7. It rose for boys too, but as we've seen several times before, the rise for boys was not concentrated so closely around the year 2012.

Often Feel Lonely

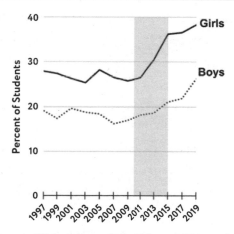

Figure 6.7. Percent of U.S. students (8th, 10th, and 12th grade) who agreed or mostly agreed with the statement "A lot of times I feel lonely." (Source: Monitoring the Future.)

This is the great irony of social media: the more you immerse yourself in it, the more lonely and depressed you become. This is true both at the individual level and at the collective level. When teens *as a whole* cut back on hanging out and doing things together in the real world, their culture changed. Their communion needs were left unsatisfied—even for those few teens who were not on social media.

After considering the four reasons that girls are particularly vulnerable, we can see why social media is a trap that ensnares more girls than boys. The lure is the promise of connecting with friends—enticing for girls who have strong needs for communion—but the reality is that girls are plunged into a strange new world in which our ancient evolved programming for real-world communities misfires continuously. Girls in virtual networks are subjected to hundreds of times more social comparison than girls had experienced for all of human evolution. They are exposed to more cruelty and bullying because social media platforms incentivize and facilitate relational aggression. Their openness and willingness to share emotions with other girls exposes them to depression and other disorders. The twisted incentive structures of social media reward the most extreme presentations of symptoms. And finally, the progress that many societies have made to reduce sexual violence and harassment in the real world is being counteracted by the facilitation of harassment and exploitation by companies that put profits above the privacy and safety of their users.

IN SUM

- Social media harms girls more than boys. Correlational studies show that heavy users of social media have higher rates of depression and other disorders than light users or nonusers. The correlation is larger and clearer for girls: Heavy users are three times as likely to be depressed as nonusers.

- Experimental studies show that social media use is a *cause*, not just a correlate, of anxiety and depression. When people are assigned to reduce or eliminate social media for three weeks or more, their mental health usually improves. Several "quasi-experiments" show that when Facebook came to campuses, or when high-speed internet came to regions and provinces, mental health declined, especially for girls and young women.

- Girls use social media a lot more than boys, and they prefer visually oriented platforms such as Instagram and TikTok, which are worse for social comparison than primarily text-based platforms such as Reddit.

- Two major categories of motivations are agency (the desire to stand out and have an effect on the world) and communion (the desire to connect and develop a sense of belonging). Boys and girls both want each of these, but there is a gender difference that emerges early in children's play: Boys choose more agency activities; girls choose more communion activities. Social media appeals to the desire for communion, but it often ends up frustrating it.

- There are at least four reasons why social media harms girls more than boys. The first is that girls are more sensitive to visual comparisons, especially when other people praise or criticize one's face and body. Visually oriented social media platforms that focus on images of oneself are ideally suited to pushing down a girl's "sociometer" (the internal gauge of where one stands in relation to others). Girls are also more likely to develop "socially prescribed perfectionism," in which a person tries to live up to impossibly high standards held by others or by society.

- The second reason is that girls' aggression is often expressed in attempts to harm the relationships and reputations of other girls, whereas boys' aggression is more likely to be expressed in physical ways. Social media has offered girls endless ways to damage other girls' relationships and reputations.

- The third reason is that girls and women more readily share emotions. When everything moved online and girls became hyperconnected, girls with anxiety or depression might have influenced many other girls to develop anxiety and depression. Girls are also more vulnerable to "sociogenic" illnesses, which means illnesses caused by social influence rather than from a biological cause.

- The fourth reason is that the internet has made it easier for men to approach and stalk girls and women and to behave badly toward them while avoiding accountability. When preteen girls open social media accounts, they are often followed and contacted by older men, and they are pressured by boys in their school to share nude photographs of themselves.

- Social media is a trap that ensnares more girls than boys. It lures people in with the promise of connection and communion, but then it multiplies the number of relationships while reducing their quality, therefore making it harder to spend time with a few close friends in real life. This may be why loneliness spiked so sharply among girls in the early 2010s, while for boys the rise was more gradual.

Chapter 7

WHAT IS HAPPENING TO BOYS?

n the important book *Stolen Focus*, the journalist Johann Hari describes
the transformation of his nine-year-old godson, a sweet child who became obsessed with Elvis Presley and who begged Hari to take him,
someday, to Elvis's home at Graceland. Hari agreed to do so. But on a visit
six years later, Hari found that the boy had changed, and was lost:

> He spent literally almost all his waking hours at home alternating
> blankly between screens—his phone, an infinite scroll of Whats-
> App and Facebook messages, and his iPad, on which he watched
> a blur of YouTube and porn. At moments, I could still see in him
> traces of the joyful little boy who sang "Viva Las Vegas," but it was
> like that person had broken into smaller, disconnected fragments.
> He struggled to stay with a topic of conversation for more than a
> few minutes without jerking back to a screen or abruptly switch-
> ing to another topic. He seemed to be whirring at the speed of
> Snapchat, somewhere where nothing still or serious could reach
> him.[1]

Hari's godson is an extreme case but not a unique one. I have heard many stories from parents about boys who fell into similar digital pits. Chris, a young man I hired to help me finalize this book, told me about his own struggles with video games and pornography that began in elementary school and have lasted to this day. He said that the combination of the two activities grew to occupy nearly every waking moment of his life, pushing out in-person play with friends, sleep, school, and later, dating. After much struggle and with the help of friends and family, Chris pulled his life together in college and found ways to moderate his gaming and pornography use. When he looks back on his years of high-intensity gaming, he remembers how much fun he had and is still grateful that gaming was a part of his life. But he is keenly aware of what he sacrificed:

> I missed out on a lot of stuff in life—a lot of socialization. I feel the effects now: meeting new people, talking to people. I feel that my interactions are not as smooth and fluid as I want. My knowledge of the world (geography, politics, etc.) is lacking. I didn't spend time having conversations or learning about sports. I often feel like a hollow operating system.

BOYS HAVE FOLLOWED A DIFFERENT PATH THROUGH THE GREAT REWIRING from girls, on average. Girls have long had higher rates of internalizing disorders than boys, and as I showed in chapter 1, that gap increased when adolescent life moved onto smartphones and social media. If we confine ourselves to examining graphs about depression, anxiety, and self-harm, we'd conclude that the Great Rewiring has been harder on girls than on boys.

And yet, if we look carefully at many of the graphs, there's plenty of evidence that boys are suffering too. Since the early 2010s, adolescent boys' rates of depression and anxiety have been rising, across many nations, though they remain at lower absolute levels than girls'. In the

No Chance of a Successful Life

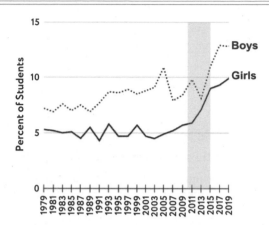

Figure 7.1. Percent of U.S. high school seniors who agreed or mostly agreed with the statement "People like me don't have much of a chance at a successful life." (Source: Monitoring the Future.)[2]

United States, the U.K., and Australia, suicide rates have been rising too, hitting both sexes, with rates always much higher for boys.[3]

There are other warning signs: Boys show evidence of pulling away from engagement with the real world well before the mental health declines of the 2010s. For boys, their time spent with friends started declining in the early 2000s, with an acceleration after 2010. Girls' rates were flatter until 2011, with a decline after that. Consider also the response to this statement: "People like me don't have much of a chance for a successful life." Only 5% of American girls used to agree with that statement back in the 1970s, and there was essentially no change until the early 2010s, as you can see in Figure 7.1. But for boys, the story is different. Their rates of agreement rose slowly, from the late 1970s through the 2000s, and then rose more quickly in the early 2010s.

In other words, the girls' story is more compact. For girls, most of the transformation in mental health takes place between 2010 and 2015, across multiple nations, and the evidence points repeatedly to the combination of smartphones and social media as major contributors to their increased anxiety and depression. For boys, in contrast, the story is more diffuse. Their decline in real-world engagement starts earlier, their mental health outcomes are more varied, and I can't point to one single technology as the primary cause of their distress. In this chapter, I'll tell a story about gradual disengagement with the real world and deepening immersion in the virtual world that reached a critical threshold when most teens got smartphones in the early 2010s, plugging them into the internet anywhere and anytime.[4] My story is more speculative than the one I told about girls in the previous chapter because we just don't know as much about what's happening to boys.

I'll tell this story using a "push-pull" analysis. First I'll show how the real world changed, since the 1970s, in ways that have made it less hospitable to boys and young men—leading many to feel more purposeless, useless, and adrift. That's the push away from the real world. Then I'll show how, starting in the 1970s and accelerating through the 2010s, the digital world brought more ways for boys to do the agency-building activities they craved, such as exploring, competing, playing at war, mastering skills, and watching increasingly hardcore pornography, all through a screen and eventually one that fit in their pockets. That's the pull.

The net effect of this push-pull is that boys have increasingly disconnected from the real world and invested their time and talents in the virtual world instead. Some boys will find career success there, because their mastery of that world can lead to lucrative jobs in the tech industry or as influencers. But for many, though it can be an escape from an increasingly inhospitable world, growing up in the virtual world makes them less likely to develop into men with the social skills and competencies to achieve success in the real world.

THE LONG DECLINE OF MALES

In 2023, Richard Reeves left his position at the Brookings Institution, where he had been a policy analyst studying economic inequality, in order to create a new organization focused on the problems of boys and men.[5] That move followed the publication of his 2022 book, *Of Boys and Men*, which laid out the evidence of a long decline in men's fortunes, achievements, and welfare in the United States since the 1970s. Part of the decline was caused by structural and economic changes that made physical strength less valuable. As America and other Western nations deindustrialized, factory work was sent to less developed nations or was done increasingly by robots. The service economy grew in its place, and women have several advantages over men, on average, in service jobs.[6]

Hanna Rosin, author of *The End of Men*, explains this transformation well: "What the economy requires now is a whole different set of skills: You need intelligence, you need an ability to sit still and focus, to communicate openly to be able to listen to people and to operate in a workplace that is much more fluid than it used to be. Those are things that women do extremely well."[7] She notes that by 2009, "for the first time in American history, the balance of the workforce tipped toward women, who continue to occupy around half of the nation's jobs."[8]

Reeves welcomes the rising fortunes of women as a desirable outcome—the natural effect of removing prior constraints on their educational and employment opportunities. For example, in 1972, women earned only 42% of bachelor's degrees. By 1982, women were just as likely as men to graduate from college. But for the next 20 years, women's enrollment rose rapidly while men's did not, so that by 2019 the gap had reversed: Women earned 59% of bachelor's degrees, while men earned just 41%.[9]

It's not just college completion. Reeves shows that at every level of education, from kindergarten through PhD, girls are leaving boys in the dust. Boys get lower grades, they have higher rates of ADHD, they are more likely to be unable to read, and they are less likely to graduate from high school, in part because they are three times as likely as girls to be expelled or suspended along the way.[10] The gender disparities are often small at the upper end, among the wealthiest families, but they grow much larger as we move down the socioeconomic ladder.

Is this a victory for girls and women? Only if you see life as a zero-sum battle between the sexes. In contrast, as Reeves puts it, "a world of floundering men is unlikely to be a world of flourishing women."[11] And the data shows that we now live in a world of floundering young men.[12]

Reeves's book helps us see the structural factors that have made it harder for boys to succeed. He describes factors such as an economy that no longer rewards physical strength, an educational system that prizes the ability to sit still and listen, and a decline in the availability of positive male role models, including fathers. After listing several such factors, Reeves adds, "The male malaise is not the result of a mass psychological breakdown, but of deep structural challenges."[13]

I think Reeves is right to focus on structural factors, but I also think there are two psychological factors missing from his story. First, the rise of safetyism in the 1980s and 1990s hit boys harder than girls, because boys engage in more rough-and-tumble play and more risky play. When playtime was shortened, pulled indoors, and over-supervised, boys lost more than girls.

The second psychological effect is the result of boys taking up online multiplayer video games in the late 2000s and smartphones in the early 2010s, both of which pulled boys decisively away from face-to-face or shoulder-to-shoulder interaction. At that point, I think we *do* see signs of a "mass psychological breakdown." Or, at least, a mass psychological *change*. Once boys had multiple internet-connected devices, many of them were swallowed whole, like Johann Hari's godson. Many boys got lost in cyberspace, which made them more fragile, fearful, and risk averse on Earth. Beginning in the late 2000s and early 2010s, American boys' rates of depression, anxiety, self-harm, and suicide began rising.[14] Boys across the Western world began showing concerning declines in their mental health.[15] By 2015, a staggering number of them said that they had no close friends, that they were lonely, and that there was no meaning or direction to their lives.[16]

BOYS WHO FAIL TO LAUNCH

Americans have long used the term "failure to launch" to describe anyone who gets off track, doesn't find employment, and ends up living back with their parents for an indefinite period of time. Young men in their late 20s are more likely to live with their parents (27% of them, in 2018), compared with young women (17%).[17] A more formal term is NEET, created by economists in the U.K. to refer to those between the ages of 16 and 24 who are Not in Education, Employment, or Training. Such young people are said to be "economically inactive." NEETs in the U.K.[18] and the United States[19] are mostly men, once you exclude all those who are disabled or who are parents caring for their own children.

American parents are more likely to say that they are worried about whether their sons will become successful adults than to say that they have that worry about their daughters.[20] Parents are also much more likely to agree with this statement about their daughters than about their sons: "Setbacks don't discourage him/her. He/she doesn't give up easily."

Parents are understandably concerned. Boys are also more vulnerable to becoming complete shut-ins, as has happened in Japan. Japanese society had long placed intense pressures on young men to succeed in school, get a prestigious job, and conform to the social expectations placed on a "salaryman." In the 1990s, when the bubble economy of the 1980s popped and the hurdles to success became higher, many young men retreated to their childhood bedrooms and shut their doors. As the economic decline made it harder for them to engage productively with the outside world, the new internet made it possible, for the first time in history, for young men to meet their agency and communion needs, to some degree, alone in their bedrooms.

These young people are called *hikikomori*, a Japanese term that means "pulling inward."[21] They live like hermits, emerging from their caves mostly at odd hours when they are less likely to see anyone, including family members. In some families, parents leave food for them by their doors. They calm their anxieties by staying inside, but the longer they stay in, the less competent they become in the outside world, fueling their anxiety about the outside world. They are trapped.

For many years, the psychiatric community treated *hikikomori* as a uniquely Japanese condition.[22] But in recent years, some young men in America and elsewhere are behaving like *hikikomori*. Some young men have even taken both the Japanese word and "NEET" as tribal identifiers. On Reddit, the r/NEET and r/hikikomori subreddits discuss everything from TV shows that espouse shut-in lifestyles to the specifics of peeing in a litter box to avoid leaving one's room.

Allie Conti at *New York* magazine spoke to one such Reddit user from North Carolina named Luca. Luca suffered from anxiety in middle school. His mother withdrew him when he was 12 and allowed him to study online from his bedroom. Boys of past generations who retreated to their bedrooms would have been confronted by boredom and almost unimaginable loneliness—conditions that would compel most homebound adolescents to change their ways or find help. Luca, however, found an online world just vivid enough to keep his mind from starving. Ten years later, he still plays video games and surfs the web all night. He sleeps all day.

Luca explains that he is not ashamed of his lifestyle. In fact, he says he's proud of it, and he contrasts it with the working lives of other young men, who let a boss tell them what to do. His room is "the opposite of a prison," he tells Conti. "It is freedom. There's no one in here but me. I can do whatever, whenever. Going outside is a prison. But this room—this room is clarity."

Luca's worldview is possible because his internet connection gives him access to a convincing simulation of many real-world pleasures— social connection, games, learning, and sex—without needing to face his anxieties and the uncomfortable uncertainty of real life. It is also representative of the swaggering and aggressive spirit that pervades the male-dominated subreddits, "chans" (such as 4chan and 8chan), and online communities (such as MGTOW—Men Going Their Own Way, without women).

BOYHOOD WITHOUT REAL-WORLD RISK

Imagine a childhood where all risk had been eliminated. Nobody ever felt the rush of adrenaline from climbing a tree when an adult had told them not to. Nobody ever experienced butterflies in their stomach as they mustered the courage to ask someone out. Picture a world where late-night outdoor adventures with friends were a thing of the past. In this childhood, there would be fewer bruises, broken bones, and broken hearts. It might sound like a safer world, but is it one you would want for your children?

Most parents would say no. And yet, somehow, something close to this is the world in which many members of Gen Z are growing up. A world with too much supervision and not enough risk is bad for all children, but it seems to be having a larger impact on boys.[23]

When I began examining boys' mental health trends for this book, I came across a striking finding. Throughout my research career, it has been common knowledge that as adolescents reach puberty, boys and girls exhibit distinct patterns in their mental health challenges. Girls typically exhibit higher rates of internalizing disorders like depression and anxiety, turning their emotions and suffering inward. Boys, on the other hand, tend to exhibit higher rates of externalizing disorders, turning their emotions outward and engaging in high risk or antisocial behavior that often affects others, such as drunk driving, violence, and drug abuse.

But around 2010, something unprecedented started happening: *Both sexes* shifted rapidly toward the pattern traditionally associated with females. There has been a notable increase in agreement with items related to internalizing disorders (such as "I feel that I can't do anything right") for both sexes, with a sharper rise among girls as you can see in figure 7.2. At the same time, agreement with items related to externalizing disorders (such as "how often have you damaged school property on purpose?") plummeted for both sexes, more sharply for boys. By 2017, boys' responses looked like those from girls in the 1990s.

One of the most widely noted traits of Gen Z is that they are not doing as much of the bad stuff that teenagers used to do. They drink less alcohol, have fewer car accidents, and get fewer speeding tickets. They have far fewer physical fights or unplanned pregnancies.[24] These are, of course, wonderful trends—nobody wants more car accidents. But because the rate of change for *so many risky behaviors* has been so rapid, I also look at these trends with concern. What if these changes came about not because Gen Z is getting wiser, but because they are withdrawing from

Internalizing and Externalizing Symptoms (U.S. Teens)

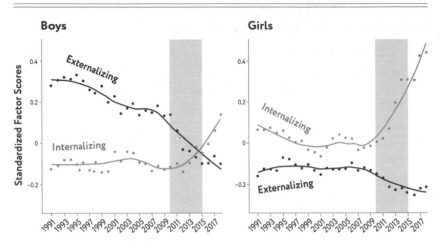

Figure 7.2. Internalizing and externalizing symptoms of U.S. high school seniors. In the 2010s, externalizing scores dropped for both sexes while internalizing scores rose. (Source: Askari et al. [2022], with data from Monitoring the Future.)[25]

the physical world? What if they are engaging in less risk-taking overall—healthy as well as unhealthy—and therefore learning less about how to manage risks in the real world?

Enjoyment of Risk-Taking

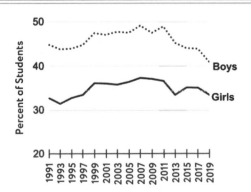

Figure 7.3. Percent of U.S. students (8th, 10th, and 12th grade) who agreed with the statement "I like to test myself every now and then by doing something a little bit risky." Enjoyment of risk-taking declined more rapidly for boys than for girls in the 2010s. (Source: Monitoring the Future.)[26]

That seems to be what happened, at least in part. Figure 7.3 shows the percentage of American students (in 8th, 10th, and 12th grades) who agreed with this statement: "I like to test myself every now and then by doing something a little bit risky."

As you can see, boys used to be much more likely to agree with that statement, and both lines in the graph hold steady in the 2000s. But then something changes. The lines decline for both genders, but the decline is steeper for boys. By 2019, boys were not far from where girls had been 10 years earlier.[27]

Hospital Admissions for Unintentional Injuries

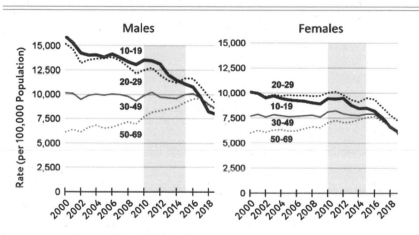

Figure 7.4. On the left: Annual rate at which U.S. males are admitted to hospitals for unintentional injuries, by age group. On the right: Same, for U.S. females. Black line is ages 10–19, the age group that used to have the highest rates of injury and now has among the lowest. (Source: Centers for Disease Control.)[28]

Boys are not just changing how they *talk or think* about risk; they are truly taking fewer risks. Figure 7.4 plots the rate of hospitalizations for unintentional injuries for four age bands, with males in the left panel and females on the right. If we look only at the years before 2010, we see that rates of hospitalization are under 10,000 per 100,000, for males as well as for females, for all age groups—with the exception of young men. Before 2010, males 10–19, along with males in their 20s, had much higher rates of hospitalization than everyone else, partly because they engaged in more risky activities and made poorer decisions.

Something changed in the 21st century. Rates of injury began to drop slowly in the 2000s, for young men only. The drop then accelerated after 2012 (and began happening for girls as well). By 2019, adolescent boys were *less* likely to be injured than adolescent girls had been in 2010. In fact, adolescent boys are now not much different from adolescent girls, or from men in their 50s and 60s.[29]

Further evidence of a change: A nationally representative study found that fall-related fractures (e.g., broken fingers and wrists) decreased slowly and steadily, among boys and girls, from 2001 to 2015, but there was one group that stood out—boys ages 10–14 showed a sharp drop after 2009. This suggests a sharp drop in doing things that could lead to a fall, such as riding bicycles over jump ramps, or climbing trees.[30]

What changed for boys? Why have they pulled away from risk in the real world? And why did these trends accelerate after 2010? We can see a possible clue in figure 7.2. The decline in boys' *externalizing* attitudes seems to have happened in two phases: a slow decline beginning in the late 1990s, with an acceleration after 2010. We see the same pattern in the decline in boys' hospitalizations for injury: slow decline in the 2000s followed by faster decline in the 2010s. We don't see that first phase of decline in either figure for girls. But the increase in *internalizing* attitudes happened only in the second phase, around the same time as girls.

So let's look at how those two phases might have affected boys in particular. How might safetyism in the 1980s and 1990s (along with males' declining societal value) and the move onto online gaming followed by smartphones in the 2000s and early 2010s have worked together to push boys away from the real world, pull them into the virtual world, and fuel their mental health crisis?

THE VIRTUAL WORLD WELCOMES BOYS

As boys began to find fewer opportunities to exercise agency, develop friendships through risky play, and pursue unsupervised adventure in the increasingly overprotective real world, they found ever increasing opportunities to build agency and friendships in the virtual world. The story starts in the 1970s with early arcade video games such as *Pong* (1972), which could later be played on one's home TV set. The first home computers arrived in the 1970s and 1980s. Computers and video games in this era, and throughout the 1990s, were of greater interest to boys than to girls.

The virtual world really began to bloom in the 1990s with the opening of the internet to the general public via web browsers such as Mosaic (1993) and AltaVista (1995), and the development of full-fledged 3-D graphics. New video game genres were developed, including "first-person shooter" games such as *Doom*, and later, massively multiplayer online games such as *RuneScape* and *World of Warcraft*.

In the 2000s, everything got faster, brighter, better, cheaper, and more private. The arrival of Wi-Fi technology increased the utility and popularity of laptop computers. Broadband high-speed internet spread quickly in this decade, making it much easier to watch videos on YouTube or Pornhub and play highly engaging online multiplayer video games on the newly released Xbox 360 (2005) and PS3 (2006). These internet-connected consoles enabled adolescents to sit alone in a room, playing for extended hours with a shifting set of strangers from around the globe. Prior to this, when boys played multiplayer video games, the other players were their friends or their siblings, sitting next to them and sharing excitement, jokes, and food.

As adolescents of both sexes got their own laptops, cell phones, and internet-connected gaming consoles, everyone was increasingly free to retreat to private spaces and do as they pleased. For boys, this opened up many new ways to satisfy their desires for agency as well as communion. In particular, it meant that boys could spend much larger amounts of time playing video games and watching pornography while alone in their bedrooms. They no longer had to use the family desktop computer or gaming console out in the living room. But did this new lifestyle— sitting alone in a room while interacting virtually—really satisfy their needs for either agency or communion?

THE VIRTUAL WORLD CONSUMES BOYS

As boys became more immersed in ever more immersive games, we see no sign of a decline in their mental health, at least until the late 2000s and early 2010s.[31] At that point, rates of suicide, depression, and anxiety all began to rise. The timing forces us to look closely at how the smartphone changed boys' tech use and engagement with the world beyond their devices. Before smartphones replaced flip phones, companies could extract children's attention only when they were seated at a computer or using a gaming console. But in the early 2010s, adolescents with smartphones became available to companies at every waking moment.

It was as if the U.S. government suddenly opened up the entire state of Alaska for drilling, and oil companies competed fiercely to stake out the best territories and start sinking wells. It is often said these days that "data is the new oil." But so is attention.

With a smartphone in every pocket, companies quickly pivoted to mobile apps, offering adolescents endless high-stimulation activities. Video game producers, pornography providers, and social media platforms adopted free-to-use, advertising-driven strategies.[32] Games also instituted pay-to-progress options—business decisions that tapped players' wallets (or parents' credit cards) directly—and got kids hooked.

Mirroring the trend seen among adolescent girls, boys' status negotiations, as well as their social and entertainment lives, increasingly moved online. Boys wandered through a bazaar of different apps, including social media, online communities, streaming platforms, gaming, pornography, and, when they got a little older, gambling and dating apps. By 2015, many boys found themselves exposed to a level of stimulation and attention extraction that had been unimaginable just 15 years earlier.

From the beginning of the digital age, the tech industry has found ever more compelling ways to help boys do the things they want to do, without having to take social and physical risks that were once needed to satisfy those desires. As traditionally "manly" skills and attributes became less valued, economically and culturally, and as the culture of safetyism grew, the virtual world stepped in to fulfill these needs directly, though not in a way that promoted skills needed for the transition to adulthood. I'll briefly discuss two of the main areas in which this happened: pornography and video games.

PORNOGRAPHY

Online hardcore pornography offers a good example of the way companies can hijack deep evolutionary drives. Evolution makes things alluring and rewarding (with a little pulse of dopamine) only when—over thousands of generations—striving for those things caused individuals to leave more surviving offspring than individuals who felt no such desire or made no such effort. Sexual attraction and mating are areas of life where evolution has left us lures and strong strivings.

In previous decades, the main way for heterosexual boys[33] to get a look at naked girls was through what we'd now consider very low-quality pornography—printed magazines that could not be sold to minors. As puberty progressed and the sex drive increased, it motivated boys to do things that were frightening and awkward, such as trying to talk to a girl, or asking a girl to dance at events organized by adults.

The internet, on the other hand, is ideally suited for the distribution of pornographic images. As data speeds increased, so did the availability of hardcore pornographic videos. Perhaps as much as 40% of all internet traffic in the late 1990s was porn.[34] The long-running Broadway play *Avenue Q*, which opened in 2003, even contained a song in which colorful puppets sang, "The internet is for porn!"

Once boys got laptops and high-speed internet, they had an infinite supply of high-quality videos showing every conceivable act, body part, and fetish, which they could watch in private, multiple times each day. A Swedish study found that 11% of boys were daily consumers in 2004 and that the number increased to 24% in 2014.[35] Another study noted that among adolescent boys who watch pornography, 59% describe it as "always stimulating"; 22% describe their use as "habitual"; 10% report that it reduces sexual interest toward potential real-life partners; and 10%

Daily Porn Users, Swedish 12th Graders

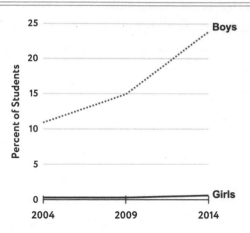

Figure 7.5. Percent of Swedish 12th graders who watch pornography "more or less daily." (Source: Donevan et al., 2022.)[36]

said it is "a kind of addiction."[37] Of course, many teen girls watch porn too, but surveys find much higher rates among boys, whether they are heterosexual or members of a sexual or gender minority.[38] When we look at daily users or users for whom porn has become an addiction that interferes with daily functioning, the male-female ratios are generally more than five or 10 to one, as in figure 7.5.

The problem is not just that modern pornography amplifies the risk for porn addiction, but that heavy porn use can lead boys to choose the easy option for sexual satisfaction (by watching porn) rather than trying to engage in the more uncertain and risky dating world. Additionally, there is evidence that heavy use can disrupt boys' and young men's romantic and sexual relationships. For example, several studies indicate that after watching porn, heterosexual men find real women less attractive, including their own partners.[39] Compulsive pornography users, who are predominantly men, are more likely to avoid sexual interactions with a partner and tend to experience lower sexual satisfaction.[40] In a 2017 meta-analysis of over 50 studies collectively including more than 50,000 participants from 10 countries, pornography consumption was "associated with lower interpersonal satisfaction outcomes in cross-sectional surveys, longitudinal surveys, and experiments." Importantly, the relationship was only significant among males.[41]

Porn separates the evolved lure (sexual pleasure) from its real-world reward (a sexual relationship), potentially making boys who are heavy users turn into men who are less able to find sex, love, intimacy, and marriage in the real world.

These trends are likely to get worse with the arrival of the metaverse, spatial video, and generative AI. Now that Meta and Apple offer headsets that let users wander through any kind of world that someone else can imagine for them, three-dimensional porn featuring "perfect" people with impossible bodies is sure to become an even stronger lure. Generative AI has already produced virtual girlfriends and boyfriends, such as CarynAI, an AI clone of a real-life 23-year-old Snapchat influencer who used thousands of hours of her YouTube content to create a sexting chatbot.[42] People are already falling madly in love with these bots as they flirt and share intimate secrets with them.[43]

As generative AI personalities improve, and as they are implanted into ever-more-lifelike sex dolls and sex robots,[44] an increasing number of heterosexual men may find that a *hikikomori* lifestyle with a programmable mechanical girlfriend is preferable to the thousands of left swipes they get on dating apps, to say nothing of the social risk of approaching a girl or woman in real life and asking her out on a date. These are the sorts of healthy risks that young men should be taking to become more competent and successful romantically.

I'm not saying that all pornography is harmful; I'm saying that immersing boys in an infinite playlist of hardcore porn videos during the sensitive period in which the sexual centers of their brains are being rewired is maybe not so good for their sexual and romantic development, or for their future partners.

VIDEO GAMES

The story around hardcore online porn is bleak, but for video games it is more complicated. When I began writing this book, I had suspected that video games would play the same role in explaining boys' problems as social media does for girls. Indeed, a meta-analysis of dozens of studies confirms that around the world men have substantially higher rates of "internet gaming disorder," while women have substantially higher rates of "social media addiction."[45] But after wading into one of the largest and most contentious areas of media research, I do not find clear evidence that would support a blanket warning to parents to keep their boys entirely away from video games.[46] The situation is different from the many studies that link girls, social media, anxiety, and depression.[47]

Unlike online pornography, researchers have found that a number of *benefits* accrue to adolescents who play video games. Some research has demonstrated that video game use is associated with increased cognitive and intellectual functioning, such as improved working memory, response inhibition, and even school competence.[48] One experiment found significant decreases in depression symptoms when an experimental group was assigned to play 30 minutes of video games three times a week for a month.[49] Other studies have found playing games cooperatively can induce players to cooperate outside the game.[50]

Nonetheless, there are at least two major harms associated with video games. First, video games can cause severe problems for a substantial subset of heavy users, like Chris, where the key is not just the *quantity* of play; it is the role that games have come to play in their lives.[51] For example, one systematic review of studies conducted during the COVID pandemic found that video game use sometimes mitigated feelings of loneliness in the short run, but it put some users into a vicious cycle because they used gaming to distract themselves from feelings of loneliness. Over time they developed a reliance on the games instead of forming long-term friendships, and this resulted in long-term stress, anxiety, and depression symptoms.[52] Building in-person relationships was, of course, difficult during COVID, but these findings are consistent with findings linking loneliness and problematic video game use from before the pandemic.[53]

Using the seven-item Gaming Addiction Scale for Adolescents, researchers have found that gamers can be divided into four groups: addicted, problematic, engaged, and casual.[54] Addicted gamers include those who admit to suffering all four of the items on a questionnaire that asks about addiction symptoms: relapse, withdrawal, conflict, and gaming causing problems. These gamers lose control over their gaming habits, as can happen with any addiction. They "become preoccupied with gaming, lie about their gaming use, lose interest in other activities just to game, withdraw from family and friends to game, and use gaming as a means of psychological escape."[55] A Canadian judge ruled in 2023 that a group of parents could sue Epic Games for the way that its game *Fortnite* addicted their sons and took over the boys' lives, leading them to skip eating, showering, and sleeping for extended periods.[56] (I note that researchers are divided on whether gaming addiction is its own disorder or if the behaviors are indicative of underlying disorders like depression or anxiety.[57])

Using the four-group framework, a "problematic gamer" endorses just two or three of those four addiction criteria. They experience negative consequences from their heavy gaming, but they don't lose control to the same degree. In contrast, "engaged gamers" play for many hours, but do not endorse any of the addictive items. Prevalence estimates vary,[58] but one 2016 study found that 1 or 2% of adult gamers qualify as having gaming addiction, 7% are problematic gamers, 4% are engaged gamers, and 87% are casual gamers.[59] Using a different set of criteria, a 2018 meta-analysis[60] found that 7% of adolescent boys can be classified as having "internet gaming disorder." That diagnosis requires "significant impairment or distress" in several aspects of a person's life.[61] (The rate for adolescent girls was estimated to be just above 1%.[62])

Different studies find different numbers, but 7% seems to be a reasonable middle-ground estimate for the percent of adolescent boys who are suffering substantial impairment in the real world (school, work, relationships) because of their heavy engagement with video games. That is one out of every 13 boys.[63]

The second major harm associated with video games is that they impose a large opportunity cost; they take up an enormous amount of time. Common Sense Media reported in 2019 (before COVID) that 41%

of adolescent boys play more than two hours per day, and 17% report playing more than four hours per day.[64] Just as with girls who devote that much time to social media, the time has to come from somewhere.[65] Those heavy gamers are missing out on sleep, exercise, and in-person social interaction with friends and family.[66] As one young man I know put it, "I really wish I had gotten to know my grandfather better before he died, instead of always playing video games when he visited."

ALL SCREENS AND NO (REAL-WORLD) PLAY

The large reduction of face-to-face social interaction is especially important for understanding what the Great Rewiring did to boys. Of course, boys are mostly playing games with other boys, so a defender of video games might argue that boys are getting *more* social interaction than they did before internet gaming, just as girls are getting more social interaction via social media. But is online gaming as good for social development as hanging out with friends in person? Or is gaming like social media, giving a lot more quantity but of much lower quality?

Video game play is happening within virtual worlds designed to maximize time spent on the platforms—just like social media. Video games are not designed to foster a small number of lasting friendships or to develop their players' social skills. As Peter Gray and other play researchers point out, one of the most beneficial parts of free play is that kids must act as legislators (who jointly make up the rules) and as judges and juries (who jointly decide what to do when rules appear to be violated). In most multiplayer video games, all of that is done by the platform. Unlike free play in the real world, most video games give no practice in the skills of self-governance.

Video games also deliver far less of the anti-phobic benefits of risky play. Video games are disembodied. They are thrilling in their own way, but they can't activate the kind of physical fear, thrill, and pounding heart that riding a roller coaster, or playing full-court basketball, or using hammers to smash things at an adventure playground can give. Jumping out of planes, having knife fights, and getting brutally murdered are just things that happen dozens of times each day for boys playing *Fortnite* or *Call of Duty*. They do not teach boys how to judge and manage risks for themselves in the real world. When video games replace real-world exploration and adventure with one's buddies, as they do for heavy users, they often produce young men who feel that they are missing something, like Chris at the opening of this chapter.

Furthermore, if video games really are a net benefit for friendships, then today's boys and young men should have more friends and be less lonely than those of 20 years ago. But in fact, the opposite is true. In 2000, 28% of 12th-grade boys reported that they often feel lonely. By 2019, that had risen to 35%. This is symptomatic of a broader "friendship recession" among men in the United States. In the 1990s, only 3% of American men reported having no close friends. By 2021, that number had risen fivefold, to 15%. A different survey in 2021 asked Americans whether there was "someone you talked to within the last six months about an important personal matter." Young men fared the worst on this question: 28% of them said no.[67] Of course, these survey questions can't prove that the arrival of online gaming in the 2000s *caused* the national increase in male loneliness, but they cast doubt on any suggestion that when boys and young men entrusted their social lives to video game companies, they entered a golden age of social connection.

Just as when girls' friendships moved onto social media platforms, boys gained quantity and lost quality. Boys thrive when they have a stable group of reliable friends, and they create their strongest and most durable friendships from being on the same team or in a stable pack, facing risks or rival teams. Virtual packs create weaker bonds, although today's increasingly lonely boys cling to them and value them because that's all they have. That's where their friends are, as Chris told me.

TECHNOLOGY, FREEDOM, AND MEANINGLESSNESS

Why, then, did boys' mental health get worse in the 2010s, just as they attained unfettered access to everything, everywhere, all the time, for free? Maybe it's because it's not healthy for *any* human being to have un-fettered access to everything, everywhere, all the time, for free.

In 1897, the French sociologist Émile Durkheim—perhaps the most profound thinker about the nature of society—wrote a book about the social causes of suicide. Drawing on data that was just becoming avail-able as governments began to keep statistics, he noted that in Europe the general rule was that the more tightly people are bound into a commu-nity that has the moral authority to restrain their desires, the less likely they are to kill themselves.

A central concept for Durkheim was *anomie*, or normlessness—an absence of stable and widely shared norms and rules. Durkheim was concerned that modernity, with its rapid and disorienting changes and its tendency to weaken the grip of traditional religions, fostered anomie and thus suicide. He wrote that when we feel the social order weakening or dissolving, we don't feel liberated; we feel lost and anxious:

> If this [binding social order] dissolves, if we no longer feel it in ex-istence and action about and above us, whatever is social in us is deprived of all objective foundation. All that remains is an artifi-cial combination of illusory images, a phantasmagoria vanishing at the least reflection; that is, nothing which can be a goal for our action.[68]

That, I believe, is what has happened to Gen Z. They are less able than any generation in history to put down roots in real-world *communities* populated by known individuals who will still be there a year later. Com-munities are the social environments in which humans, and human childhood, evolved. In contrast, children growing up after the Great Re-wiring skip through multiple *networks* whose nodes are a mix of known and unknown people, some using aliases and avatars, many of whom will have vanished by next year, or perhaps by tomorrow. Life in these networks is often a daily tornado of memes, fads, and ephemeral micro-dramas, played out among a rotating cast of millions of bit players. They

Life Often Feels Meaningless

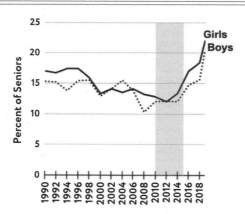

Figure 7.6. Percent of U.S. high school seniors who agreed or mostly agreed with the statement "Life often feels meaningless." (Source: Monitoring the Future.)

have no roots to anchor them or nourish them; they have no clear set of norms to constrain them and guide them on the path to adulthood.

Durkheim and his concept of anomie can explain why all of a sudden, in the early 2010s, boys as well as girls began to agree much more vigorously with the statement "Life often feels meaningless."

Boys and girls have taken different paths through the Great Rewiring, yet somehow they have ended up in the same pit, where many are drowning in anomie and despair. It is very difficult to construct a meaningful life on one's own, drifting through multiple disembodied networks. Like Johann Hari's godson, their consciousness ends up "broken into smaller, disconnected fragments." Human children and human bodies need to be rooted in human communities. Children must grow up on Earth before we can send them to Mars.

IN SUM

- Like girls, boys got more depressed and anxious in the early 2010s, in many countries. Unlike girls, boys experienced a slow decline since the 1970s in achievement and engagement in school, work, and family life.

- Boys and young men withdrew much of their time and effort from the physical world (which was increasingly opposed to unsupervised play, exploration, and risk-taking) and invested it in the rapidly expanding virtual world.

- Boys are at greater risk than girls of "failure to launch." They are more likely to become young adults who are "Not in Education, Employment, or Training." Some Japanese men developed an extreme form of lifelong withdrawal to their bedrooms; they are called *hikikomori*.

- In the early 2010s, American teen boys' thinking patterns shifted from what they had traditionally been (higher rates of externalizing cognitions and behaviors than internalizing) to a pattern more commonly shown by girls (higher rates of internalizing). At the same time, boys also began to shun risk (more so than happened for girls).

- As boys engaged in fewer risky activities outdoors or away from home, and began spending more time at home on screens, their mental health did not decline in the 1990s and 2000s. But something changed in the early 2010s, and their mental health then began to decline.

- Once boys got smartphones, they—like girls—moved even more of their social lives online, and their mental health declined.

- One way that smartphones—amplified by high-speed internet— have affected boys' lives is by providing unlimited, free, hardcore pornography accessible anytime, anywhere. Porn is an example of how tech companies have made it easy for boys to satisfy powerful evolved desires without having to develop any skills that would help them make the transition to adulthood.

- Video games offer boys and girls a number of benefits, but there are also harms, especially for the subset of boys (in the ballpark of 7%) who end up as problematic or addicted users. For them, video games do seem to cause declining mental and physical health, family strife, and difficulties in other areas of life.

- As with social media for girls, spending hours "connecting" with others online produces an increase in the quantity of social interactions and a decrease in the quality of social relationships. Boys, like girls, became lonelier during the Great Rewiring. Some boys use video games to strengthen their real-world packs, but for many others, video games made it easier for them to retreat to their bedrooms rather than doing the hard work of maturing in the real world.

- The Great Rewiring of Childhood pulled young people out of real-world communities, including their own families, and created a new kind of childhood lived in multiple rapidly shifting networks. One inevitable result was anomie, or normlessness, because stable and binding moralities cannot form when everything is in flux, including the members of the network.

- As the sociologist Émile Durkheim showed, anomie breeds despair and suicide. This may be why boys and girls, who followed different paths through the Great Rewiring, ended up in the same place, with a sudden and rapid increase in the feeling that their lives were meaningless.

Chapter 8

SPIRITUAL ELEVATION
AND DEGRADATION

In the previous three chapters, I described a great deal of research on harms to children and adolescents caused by the phone-based childhood. But now I'd like to write less as a social scientist than as a fellow human being who has felt overwhelmed, personally and perpetually, since around 2014. It feels as if something very deep changed in the 2010s. On college campuses, there seemed to be a shift from discover mode to defend mode. In American politics, things got even stranger. I've been struggling to figure out: What is happening to us? How is technology changing us? Most of my research since then has been an effort to answer those questions. Along the way I have found inspiration and insight from an eclectic set of academic sources and from several ancient traditions. I think I can best convey what is happening to us by using a word rarely used in the social sciences: *spirituality*. The phone-based life produces spiritual degradation, not just in adolescents, but in all of us.

In *The Happiness Hypothesis*, I wrote a chapter titled "Divinity with or Without God," in which I presented my research on the moral emotions, including disgust, moral elevation, and awe. I showed that people perceive

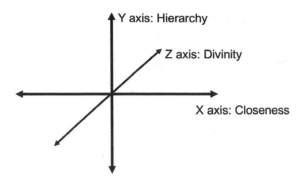

Figure 8.1. Three dimensions of social space.

three dimensions of social space. In every society, you'll find that people distinguish between those they feel close to and those who are more distant; that's the horizontal dimension, the *x* axis in figure 8.1. Then there are those who are higher in rank or social status and who are owed deference by those who are lower. That's the vertical dimension of hierarchy, the *y* axis. Many languages force people to mark those two dimensions when they speak, as in French when one must decide whether to address someone as *vous* or as *tu*.

But there's another vertical dimension, shown as the *z* axis coming out of the page. I called it the divinity axis because so many cultures wrote explicitly that virtuous actions bring one upward, closer to God, while base, selfish, or disgusting actions bring one downward, away from God and sometimes toward an anti-divinity such as the Devil. Whether or not God exists, people simply *do* perceive some people, places, actions, and objects to be sacred, pure, and elevating; other people, places, actions, and objects are disgusting, impure, and degrading (meaning, literally, "brought down a step").

Thomas Jefferson offered a secular description of the *z* axis in 1771. In a letter advising a relative on what books to buy for his library, Jefferson urged the inclusion of novels and plays. He justified his advice by reflecting on the feelings one gets from great literature:

> When any . . . act of charity or of gratitude, for instance, is presented either to our sight or imagination, we are deeply impressed

with its beauty and feel a strong desire in ourselves of doing charitable and grateful acts also. On the contrary, when we see or read
of any atrocious deed, we are disgusted with its deformity, and
conceive an abhorrence of vice.

Jefferson specifically described moral elevation as the opposite of
disgust. He then considered the example of a contemporary French play,
and asked whether the virtues of fidelity and generosity exemplified by
its hero do not

> dilate [the reader's] breast and elevate his sentiments as much as
> any similar incident which real history can furnish? Does [the
> reader] not in fact feel himself a better man while reading them,
> and privately covenant to copy the fair example?

Jefferson's use of the word "elevate" captures for all of us the feeling
that we are lifted "up" in some way. Conversely, witnessing people behaving in petty, nasty ways, or doing physically disgusting things, triggers
revulsion. We feel pulled "down" in some way. We close off and turn away.
Such actions are incompatible with our elevated nature. This is how I'm
using the word "spiritual." It means that one endeavors to live more of
one's life well above zero on the z axis. Christians ask, "What would Jesus
do?" Secular people can think of their own moral exemplar. (I should
point out that I am an atheist, but I find that I sometimes need words and
concepts from religion to understand the experience of life as a human
being. This is one of those times.)

So now I want to ask: Does the phone-based life generally pull us upward or downward on this vertical dimension? If it is downward, then
there is a cost even for those who are not anxious or depressed. If it is
downward, then there is *spiritual* harm, for adults as well as for adolescents, even for those who think that their mental health is fine. There
would also be harm to society if more people are spending more time
below zero on the z axis. We would perceive a general society-wide degradation that would be hard to put into words.

In the rest of this chapter, I'll draw on wisdom from ancient traditions and modern psychology to try to make sense of how the phone-based life affects people spiritually by blocking or counteracting six spiritual practices: shared sacredness; embodiment; stillness, silence, and focus; self-transcendence; being slow to anger, quick to forgive; and finding awe in nature.

SPIRITUAL PRACTICES

The social psychologist David DeSteno published a book in 2021 with the provocative title *How God Works: The Science Behind the Benefits of Religion*.[1] In his book, DeSteno reviewed psychological research on the efficacy of spiritual practices such as meditation, prayer, confession, and atonement rituals. Though researchers have not found evidence that prayer works to change outcomes in the world, such as curing a child of cancer, DeSteno found that there *is* abundant evidence that keeping up certain spiritual practices improves well-being. The mechanism often involves reducing self-focus and selfishness, which prepares a person to merge with or be open to something beyond the self. When communities engage in these practices together, and especially when they move together in synchrony, they increase cohesion and trust, which means that they also reduce anomie and loneliness.[2]

Looking at these six practices can help us see what many of us have lost as we have entwined our lives more fully with our digital assistants. These practices point us to ways to improve our own lives, and those of children and adolescents too. These are practices that all of us can do, whether we are religious or not, to flourish and connect in our age of anxiety and fragmentation. In fact, they may be more important for those who are *not* religious and don't get exposure to these practices inside a faith community.

1. Shared Sacredness

Durkheim argued that *Homo sapiens* could just as well be called *Homo duplex*, or two-level man, for we exist on two very different levels. We spend most of our lives as individuals pursuing our own interests. He called this the realm of the "profane," which means the ordinary day-to-day world where we are very concerned about our own wealth, health, and reputation. But Durkheim showed that nearly all societies have created rituals and communal practices for pulling people "up," temporarily, into the realm of the sacred, where the self recedes and collective interests predominate. Think of Christians singing hymns together every Sunday in church; think of Muslims circling the Kaaba in Mecca; think of civil rights marchers singing as they walked. Evidence that these two levels are available to everyone, even outside a religious context, can be found in the ways that fans of sports teams use similar techniques to bind themselves together before a game with pep rallies, the singing of fight songs, and shared consciousness alteration (usually from alcohol), along with a variety of quasi-religious rituals, superstitions, and body markings.[3] It is a thrill to be one of thousands of fans in a stadium, all singing and stomping in unison after each goal or touchdown. Durkheim called this state of energized communion "collective effervescence."

This is one of the founding insights of sociology: Strong communities don't just magically appear whenever people congregate and communicate. The strongest and most satisfying communities come into being when something lifts people out of the lower level so that they have powerful collective experiences. They all enter the realm of the sacred together, at the same time. When they return to the profane level, where they need to be most of the time to address the necessities of life, they have greater trust and affection for each other as a result of their time together in the sacred realm. They are also happier and have lower rates of suicide. In contrast, transient networks of disembodied users, interacting asynchronously, just can't cohere the way human communities have from time immemorial. People who live only in networks, rather than communities, are less likely to thrive.

To enable their adherents to share collective experiences, religions mark off certain *times* (such as the Sabbath and holy days), *places* (shrines, churches, temples), and *objects* (the cross, the Bible, the Qur'an) as sacred.

They are separate from the profane world; the faithful must protect them from desecration. The Hebrew word for holiness (*kadosh*) literally means "set apart," or "separated."

But what happens when social life becomes virtual and everyone interacts through screens? Everything collapses into an undifferentiated blur. There is no consensual space—at least not any kind that feels real to human minds that evolved to navigate the three dimensions of planet Earth. In the virtual world, there is no daily, weekly, or annual calendar that structures when people can and cannot do things. Nothing ever closes, so everyone acts on their own schedule.[4]

In short, there is no consensual structuring of time, space, or objects around which people can use their ancient programming for sacredness to create religious or quasi-religious communities. Everything is available to every individual, all the time, with little or no effort. There is no Sabbath and there are no holy days. Everything is profane. Living in a world of structureless anomie makes adolescents more vulnerable to online recruitment into radical political movements that offer moral clarity and a moral community, thereby pulling them further away from their in-person communities.

We could create healthier environments for ourselves and our children if we could reconnect with the rhythms of the calendar and of our communities. This might include taking part in regular religious services or joining other groups organized for a moral, charitable, or spiritual purpose. It could include establishing family rituals such as a digital Sabbath (one day per week with reduced or no digital technology, combined with enjoyable in-person activities) or marking holidays together consistently, ideally with other families. All such practices would endow time and space with some of the social meaning they have lost.

2. Embodiment

Once time and space are structured for sacredness, rituals can proceed, and rituals require bodies in motion. Prayer or meditation can be silent and motionless, but religions usually prescribe some kind of movement that marks the activity as devotional and adds to its symbolism. Christians kneel, Muslims prostrate themselves toward Mecca, Sufis have "whirling dervishes," and Jews "daven," which involves praying out loud while moving or rocking in a certain way. Congregations sing and dance together, which opens their hearts to each other and to God.[5] DeSteno notes that synchronous movement during religious rituals is not only very common; it is also an experimentally validated technique for enhancing feelings of communion, similarity, and trust, which means it makes a group of disparate individuals feel as though they have merged into one.[6]

Anyone who participated in a Zoom-based wedding, funeral, or religious service during the COVID pandemic knows how much is lost when rituals go virtual. Humans evolved to be religious by being together and moving together. The Great Rewiring reduced synchronous physical movement—indeed all physical movement—and then COVID lockdowns reduced it further still.

Perhaps the most important embodied activity that binds people together is eating. Most major holy days and rites of passage involve a feast, or at least a shared meal, often with foods specific for that day or ritual. Imagine how you'd feel if you were an American and someone in your family said on Thanksgiving that he was feeling hungry so he was going to take his portion of turkey, stuffing, and cranberry sauce now, an hour before dinner, and eat it alone in another room. Then he'd come back and sit with the family while they ate. No. The assembled family and friends must share the food, and this is among the most widespread of human customs: People who "break bread" together have a bond.[7] The simple act of eating together, especially from the same plate or serving dish, strengthens that bond and reduces the likelihood of conflict. This is one deficiency the virtual world can never overcome, no matter how good VR gets.

Many spiritual practices are amplified by bodies in motion and in proximity. When everything is done on a screen, and perhaps done alone in your bedroom, you cannot activate the neural circuits that evolved along with spiritual practice,[8] so it is much more difficult to enter Durkheim's realm of the sacred. A healthier way to live would be to seek out more in-person communal events, especially those that feel as though there is an elevated or moral purpose and that involve some synchronous movement, such as religious services, or live concerts for some musicians with devoted followings. Especially in the years after COVID, many of us would benefit from changing habits adopted during the pandemic and not always choosing the easy, remote option.

Sports are not exactly spiritual, but playing them depends on some of spirituality's key ingredients for bonding people together, like coordinated and collective physical movement and group celebrations. Research consistently shows that teens who play team sports are happier than those who don't.[9] Humans are embodied; a phone-based life is not. Screens lead us to forget that our physical bodies matter.

3. Stillness, Silence, and Focus

Bodies aren't always in motion during spiritual practices; some practices use stillness, although even being still is physically intense. Meditation traditions prescribe how to sit, breathe, and visualize the body. The Buddha followed the "eightfold noble path" to enlightenment. The eighth element, interacting with all the others, is *samadhi*, often translated as "meditative absorption." Without training, the mind flits around like a jumping monkey. With our multiscreen, multitasking lives, the monkey jumps even more frantically, as with Johann Hari's godson. One of the fundamental teachings of the Buddha is that we can train our minds.

Meditation helps to calm the monkey mind. Over time, the nature of conscious experience changes, even when one is not meditating. Studies on Buddhist monks suggest that their intense meditation practices alter their brains in lasting ways, decreasing activation in brain areas related to fear and negative emotionality. That's a sign that they have come to live in the openness of discover mode, rather than in the guardedness of defend mode.[10]

This is why many religions have monasteries and monks. Those seeking spiritual growth are well served by separating themselves from the noise and complexity of human interactions, with their incessant words and profane concerns. When people practice silence in the company of equally silent companions, they promote quiet reflection and inner work, which confers mental health benefits. Focusing your attention and meditating have been found to reduce depression and anxiety.[11] You don't need to become a monk or join a monastery; many ordinary people gain these benefits by taking a vow of silence for a day, a week, or more as they join with others on meditation retreats. Even brief sessions of mindfulness meditation—10 minutes each day—have been found to reduce irritability, negative emotions, and stress from external pressures.[12] In fact, mindfulness practices, originating in the spiritual realm, have now been routinely introduced into psychiatric and medical practice with growing empirical evidence to support their efficacy.[13]

The Buddha described *samadhi* as a state of "mental unity." He said, "When you gain *samadhi*, the mind is not scattered, just as those who protect themselves from floods guard the levee."[14] Smartphones and social media smash the levee, flood consciousness with alerts and triviality, fill the ears with sounds, fragment attention, and scatter consciousness.[15] The phone-based life makes it difficult for people to be fully present with others when they are with others, and to sit silently with themselves when they are alone. If we want to experience stillness and silence, and if we want to develop focus and a sense of unified consciousness, we must reduce the flow of stimulation into our eyes and ears. We must find ample opportunities to sit quietly, whether that is in meditation,[16] or by spending more time in nature, or just by looking out a car window and thinking on a long drive, rather than always listening to something, or (for children in the back seat) watching videos the whole way.

4. Transcending the Self

Think about your last spiritual experience, perhaps a moment of awe in nature, or a moment of moral elevation or inspiration from witnessing an act of moral beauty. Did you feel more self-conscious at that moment, or less?

Self-transcendence is among the central features of spiritual experience, and it turns out that the loss of self has a neural signature. There is a set of linked structures in the brain that are more active whenever we are processing events from an egocentric point of view—thinking about what *I* want, what *I* need to do next, or what other people think of *me*. These brain structures are so often active together that they are collectively called the default mode network (DMN), meaning it is what the brain is usually doing, except in the special times when it is not.[17]

We might call it the profane mode network. Research has found that the DMN is less active when people engage in a variety of spiritual practices, including meditation, prayer, and the use of psychedelic drugs (in supportive settings) such as psilocybin, which is the class of drugs the world's indigenous religions most widely use.[18] In his 2023 book, *Awe*, the social psychologist Dacher Keltner wrote,

> As our default self vanishes, other studies have shown, awe shifts us from a competitive, dog-eat-dog mindset to perceive that we are part of networks of more interdependent, collaborating individuals. We sense that we are part of a chapter in the history of a family, a community, a culture. An ecosystem.[19]

When the DMN is quieter, we are better able to deeply connect to something beyond ourselves. What does social media do to the DMN? A social media "platform" is, almost by definition, a place that is all about you. You stand on the platform and post content to influence how others perceive you. It is almost perfectly designed to crank up the DMN to maximum and pin it there. That's not healthy for any of us, and it's even worse for adolescents.[20]

The Buddhist and Taoist traditions wrote extensively about the obstacles our egos throw up on the path to enlightenment. Our consciousness gets jerked around by profane concerns. In the Tao Te Ching, a foundational Taoist text from the fourth century BCE, we find this:

Bedevilments arising in the mind are ideas of self and others, ideas of glory and ignominy, ideas of gain and loss, ideas of right and wrong, ideas of profit and honor, ideas of superiority. These are dust on the pedestal of the spirit, preventing freedom.

Social media is a fountain of bedevilments. It trains people to think in ways that are exactly contrary to the world's wisdom traditions: *Think about yourself first; be materialistic, judgmental, boastful, and petty; seek glory as quantified by likes and followers.* Many users may believe that the implicit carrots and sticks built into platforms like Instagram don't affect them, but it's hard not to be affected unconsciously. Unfortunately, most young people become heavy users of social media during the sensitive period for cultural learning, which runs from roughly age 9 to 15.[21]

To experience more self-transcendence, we need to turn down the things in our lives that activate the profane mode network and bind us tightly to our egos, such as time on social media. We need to seek out conditions and activities that have the opposite effect, as most spiritual practices do, including prayer, meditation, mindfulness, and for some people psychedelic drugs, which are increasingly found to be effective treatments for anxiety and depression.[22]

5. Be Slow to Anger, Quick to Forgive

The Tao Te Ching lists "ideas of right and wrong" as a bedevilment. In my 35 years of studying moral psychology, I have come to see this as one of humanity's greatest problems: We are too quick to anger and too slow to forgive. We are also hypocrites who judge others harshly while automatically justifying our own bad behavior. As Jesus said in the Sermon on the Mount,

Do not judge, so that you may not be judged. For the judgment you give will be the judgment you get, and the measure you give will be the measure you get.[23]

Jesus was not telling us to avoid judging others entirely; he was warning us to judge thoughtfully, and to beware of using different standards for others than we use for ourselves. In the next verse he says, "Why do you see the speck in your neighbor's eye but do not notice the log in your own eye?"[24] He urges us to fix ourselves first, before we criticize anyone else.

Social media trains us to do the opposite. It encourages us to make rapid public judgments with little concern for the humanity of those we criticize, no knowledge of the context in which they acted, and no awareness that we have often done the very thing for which we are publicly shaming them.

The Buddhist and Hindu traditions go even further, urging us to forswear judgment entirely. Here is one of the deepest insights ever attained into the psychology of morality, from the eighth-century Chinese Zen master Seng-ts'an:

The Perfect Way is only difficult
for those who pick and choose;
Do not like, do not dislike;
all will then be clear.
Make a hairbreadth difference,
and Heaven and Earth are set apart;
If you want the truth to stand clear before you,
never be for or against.
The struggle between "for" and "against"
is the mind's worst disease.[25]

We can't follow Seng-ts'an's advice literally; we can't avoid making moral distinctions and judgments entirely. (Indeed, monotheistic religions are full of moral distinctions and judgments.) But I believe his point was that the mind, left to its own devices, evaluates everything immediately, which shapes what we think next, making it harder for us to find the truth. This insight is the foundation of the first principle of moral psychology, which I laid out in *The Righteous Mind*: Intuitions come first, strategic reasoning second. In other words, we have an immediate gut feeling about an event, and then we make up a story after the fact to justify our rapid judgment—often a story that paints us in a good light.

The world's major religions advise us to turn down the judgmentalism and turn up the forgiveness. In the Torah, God commands the Israelites, "Do not seek revenge or bear a grudge against anyone among your people, but love your neighbor as yourself."[26] Thousands of years later, Martin Luther King Jr. used the power of forgiveness, as developed in the Judeo-Christian tradition, to inspire those in the civil rights movement to act in elevated ways that would win hearts and minds:

> We must develop and maintain the capacity to forgive. He who is devoid of the power to forgive is devoid of the power to love. There is some good in the worst of us and some evil in the best of us. When we discover this, we are less prone to hate our enemies.[27]

Of course, religion has at times motivated people to be cruel, racist, and genocidal. Religious practitioners, like all people, are often hypocrites. Nonetheless, religious injunctions to be slower to judge and quicker to forgive are good for maintaining relationships and improving mental health. Social media trains people to do the opposite: Judge quickly and publicly, lest ye be judged for not judging whoever it is that we are all condemning today. Don't forgive, or your team will attack you as a traitor.

From a spiritual perspective, social media is a disease of the mind. Spiritual practices and virtues, such as forgiveness, grace, and love, are a cure. As Buddha put it:

> In this world, hate never yet dispelled hate.
> Only love dispels hate. This is the law,
> Ancient and inexhaustible.
> You too shall pass away.
> Knowing this, how can you quarrel?[28]

6. Find Awe in Nature

It is impossible to overstate the role that the grandeur of nature has played in human spirituality. Psalm 19 sings out, "The heavens are telling the glory of God, and the firmament proclaims his handiwork." Here is Ralph Waldo Emerson in 1836 describing the effects of that handiwork when he walks in a forest:

> In the woods ... these plantations of God, a decorum and sanctity reign. . . . Standing on the bare ground, my head bathed by the blithe air and uplifted into infinite space, all mean egotism vanishes. I become a transparent eyeball; I am nothing; I see all; the currents of the Universal Being circulate through me; I am part or parcel of God.[29]

In 2003, Dacher Keltner and I published a review paper on the emotion of awe in which we argued that awe is triggered by two simultaneous perceptions: first, that what you are looking at is vast in some way, and, second, that you can't fit it into your existing mental structures.[30] That combination seems to trigger a feeling in people of being small in a profoundly pleasurable—although sometimes also fearful—way. Awe opens us to changing our beliefs, allegiances, and behaviors.

Dacher went on to become the preeminent scientist studying awe. He and his students collected thousands of accounts of awe experiences from people around the world and sorted them into the eight most common categories, which he calls the "eight wonders of life." They are moral beauty, collective effervescence, nature, music, visual design, spiritual and religious awe, life and death, and epiphanies (moments in which a new and grand understanding dawns).

Awe can be triggered in many ways, but the beauty of nature is among the most reliable and accessible methods. After hearing Dacher in a podcast conversation[31] describe the "awe walks" he took while grieving his brother's death from cancer, I decided to add a session on awe and beauty to the undergraduate Flourishing class that I teach at New York University. I told my students to listen to the podcast and then take a walk, slowly, anywhere outside, during which they must not take out their phones. The written reflections they turned in for that week's homework were among the most beautiful I've seen in my 30 years as a professor.

Some students simply walked slowly through the streets of Greenwich Village, around NYU, noticing for the first time the architectural flourishes (visual design) on 19th-century buildings that they had passed many times. But the most powerful reports came from those who walked through parks. One student, Yi-Mei, began her awe walk in Washington Square Park, which is the green heart of the NYU campus. It was a perfect April day when the cherry trees were in full bloom:

> I was so overwhelmed with how beautiful the park seemed in the spring that I took time sitting on a bench contemplating its beauty and finding moral delight and affection toward the people that I see walking around, smiling at each of them as they look at me.

She was so inspired by this new experience in a familiar park that she then went up to New York's Central Park, which she did not need to do to complete the assignment. There she was dazzled by the reflections of sunlight on a small lake, "as if there were sparkles sprinkled on it as decoration, and on the trees. To me, it was as if everything came to life."

Several students wrote that before their awe walks, they rarely took the time to absorb the beauty of the world around them. Washington Square Park is among the most beautiful urban parks in the United States, and NYU students walk through it often, yet many hadn't truly seen it before.

Many of the students suffer from anxiety, and several found that natural beauty was an effective treatment. Here is Yi-Mei again:

> It felt as if the experience of beauty and awe made me more generous and drawn into the present. The petty concerns of the past

suddenly felt dull, and to worry about the future felt unnecessary because of how secure and calm I felt now. It was like I was experiencing a stretch of time and saying to myself and my anxiety that "everything will be OK." There was also a swarming feeling of happiness and simply wanting to connect with and talk to people.

In a 2023 review paper, Dacher and a colleague listed five ways that awe improves well-being: Awe causes "shifts in neurophysiology, a diminished focus on the self, increased prosocial relationality, greater social integration, and a heightened sense of meaning."[32] Yi-Mei experienced them all during her quiet walk through two parks.

Humans evolved in nature. Our sense of beauty evolved to attract us to environments in which our ancestors thrived, such as grasslands with trees and water, where herbivores are plentiful, or the ocean's edge, with its rich marine resources. The great evolutionary biologist E. O. Wilson said that humans are "biophilic," by which he meant that humans have "the urge to affiliate with other forms of life."[33] This is why people travel to wondrous natural destinations. It's why the great landscape architect Frederick Law Olmsted designed Central Park the way it is, with fields, woods, lakes, and a small zoo where my children delighted in feeding sheep and goats. It's why children love to explore the woods and turn over rocks, to see what they'll find crawling underneath. This is also why spending time in beautiful natural settings reduces anxiety in those suffering from anxiety disorders.[34] It's like coming back home.

Yet one of the hallmarks of the Great Rewiring is that children and adolescents now spend far less time outside, and when they are outside, they are often looking at or thinking about their phones. If they encounter something beautiful, such as sunlight reflected on water, or cherry blossoms wafting on gentle spring breezes, their first instinct is to take a photograph or video, perhaps to post somewhere. Few are open to losing themselves in the moment as Yi-Mei did.

One can certainly feel some kinds of awe while using a smartphone. Indeed, you can watch endless YouTube videos about people who performed heroic deeds (moral beauty). You can find the most extraordinary photos and videos ever taken of the world's most beautiful places. These experiences are valuable. But as we've seen before, our phones drown us in quantity while reducing quality. You watch a morally elevating short video, feel moved, and then scroll to the next short video, in which someone is angry about something. You see a photo of Victoria Falls, taken from a drone that gives you a better view than you could ever get in person, and yet, because the entire image is displayed on a screen the size of your hand, and because you did no work to get to the falls, it's just not going to trigger as much awe as you'd get from hiking up to a much smaller waterfall yourself.

If we want awe to play a larger and more beneficial role in our lives, we need to make space for it. As a result of doing my own awe walk the same week my students did them, I now take my AirPods out of my ears when I'm walking in any park or natural setting. I no longer try to cram in as many audiobooks and podcasts, at 1.5 times normal speed, as my brain can receive. As for our children, if we want awe and natural beauty to play a larger role in their lives, we need to make deliberate efforts to bring them or send them to beautiful natural areas. Without phones.

THE GOD-SHAPED HOLE

Soon before his death in 1662, the French philosopher Blaise Pascal wrote a paragraph often paraphrased as "there is a God-shaped hole in every human heart."[35] I believe he was right. In *The Righteous Mind*, I drew on the writings of Charles Darwin and the biologist David Sloan Wilson[36] to explain how natural selection might have carved that hole: Humanity went through a long period of what is known as *multilevel selection* in which groups competed with groups, while at the same time individuals competed with individuals within each group. The most cohesive groups won, and humans evolved—by both biological and cultural evolution— an adaptation that made their groups even more cohesive: religiosity (including both the fear and the love of gods).

Many of my religious friends disagree about the origin of our God-shaped hole; they believe that the hole is there because we are God's creations and we long for our creator. But although we disagree about its origins, we agree about its implications: There is a hole, an emptiness in us all, that we strive to fill. If it doesn't get filled with something noble and elevated, modern society will quickly pump it full of garbage. That has been true since the beginning of the age of mass media, but the garbage pump got 100 times more powerful in the 2010s.

It matters what we expose ourselves to. On this the ancients universally agree. Here is Buddha: "We are what we think. All that we are arises with our thoughts."[37] And here is Marcus Aurelius: "The things you think about determine the quality of your mind. Your soul takes on the color of your thoughts."[38]

In a phone-based life, we are exposed to an extraordinary amount of content, much of it chosen by algorithms and pushed to us via notifications that interrupt whatever we were doing. It's too much, and a lot of it pulls us downward on the divinity dimension. If we want to spend most of our lives above zero on that dimension, we need to take back control of our inputs. We need to take back control of our lives.

IN SUM

- When people see morally beautiful actions, they feel as though they have been lifted up—elevated on a vertical dimension that can be labeled divinity. When people see morally repulsive actions, they feel as though they have been pulled downward, or degraded.

- A phone-based life generally pulls people downward. It changes the way we think, feel, judge, and relate to others. It is incompatible with many of the behaviors that religious and spiritual communities practice, some of which have been shown to improve happiness, well-being, trust, and group cohesion, according to researchers such as David DeSteno. I described six such practices.

- First, Émile Durkheim showed that human beings move up and down between two levels: the profane and the sacred. The profane is our ordinary self-focused consciousness. The sacred is the realm of the collective. Groups of individuals become a cohesive community when they engage in rituals that move them in and out of the realm of the sacred together. The virtual world, in contrast, gives no structure to time or space and is entirely profane. This is one reason why virtual communities are not usually as satisfying or meaning-giving as real-world communities.

- Second, religious rituals always involve bodily movement with symbolic significance, often carried out synchronously with others. Eating together has a special power to bond people together. The virtual world is, by definition, disembodied, and most of its activities are conducted asynchronously.

- Third, many religions and spiritual practices use stillness, silence, and meditation to calm the "jumping monkey" of ordinary consciousness and open the heart to others, God, or enlightenment. Meditation has been shown to promote well-being, even brief regular meditation in fully secular contexts. The phone-based life, in contrast, is a never-ending series of notifications, alerts, and distractions, fragmenting consciousness and training us to fill every moment of consciousness with something from our phones.

- Fourth, a defining feature of spirituality is self-transcendence. There is a network of brain structures (the default mode network) that becomes less active during moments of self-transcendence, as if it were the neural basis of profane consciousness. Social media keeps the focus on the self, self-presentation, branding, and social standing. It is almost perfectly designed to prevent self-transcendence.

- Fifth, most religions urge us to be less judgmental, but social media encourages us to offer evaluations of others at a rate never before possible in human history. Religions advise us to be slower to anger and quicker to forgive, but social media encourages the opposite.

- Sixth, the grandeur of nature is among the most universal and easily accessible routes to experiencing awe, an emotion that is closely linked to spiritual practices and progress. A simple walk in a natural setting can cause self-transcendence, especially if one pays close attention and is not attending to a phone. Awe in nature may be especially valuable for Gen Z because it counteracts the anxiety and self-consciousness caused by a phone-based childhood.

- There is a "God-shaped hole" in every human heart. Or, at least, many people feel a yearning for meaning, connection, and spiritual elevation. A phone-based life often fills that hole with trivial and degrading content. The ancients advised us to be more deliberate in choosing what we expose ourselves to.

This concludes part 3, in which I laid out the harms of a phone-based childhood (and of a phone-based adulthood). Now we're ready to talk about what we can do about it, in part 4. I'll show that we can change things if we act together.

Part 4

COLLECTIVE ACTION
FOR HEALTHIER
CHILDHOOD

Chapter 9

PREPARING FOR
COLLECTIVE ACTION

The most common response I get when I say that we need to delay the age at which children get smartphones and social media accounts is "I agree with you, but it's too late." It has become so ordinary for 11-year-olds to walk around staring down at their phones, swiping through bottomless feeds, that many people cannot imagine that we could change it if we wanted to. "That ship has sailed," they tell me, or "that train has left the station." But to me, these transportation metaphors imply that we need to act right away. I've been on airplanes that left the gate and were then called back when a safety issue was discovered. After the *Titanic* sank in 1912, its two sister ships were pulled out of service and modified to make them safer. When new consumer products are found to be dangerous, especially for children, we recall them and keep them off the market until the manufacturer corrects the design.

In 2010, teens, parents, schools, and even tech companies didn't know that smartphones and social media had so many harmful effects. Now we do. In 2010 there was little sign of a mental health crisis. Now it's all around us.

We are not helpless, although it often feels that way because smartphones, social media, market forces, and social influence combine to pull us into a trap. Each of us, acting alone, perceives that it's too difficult or costly to do the right thing. But if we can act together, the costs go way down.

In this short chapter, I explain what collective action problems are and I describe some of the common mechanisms used to solve them. Then, in each of the remaining chapters of part 4, I'll show what governments, tech companies, schools, and parents can do to reverse the disastrous transition from play-based to phone-based childhood.

COLLECTIVE ACTION PROBLEMS

Social scientists have long studied traps where each individual does what she thinks is best for herself (such as overfishing in a local pond), even though, when everyone makes the same choice, it leads to a bad outcome for all (the pond stops producing any fish). If the group could coordinate (such as by setting a limit on how many fish each resident can take), the long-term outcome would be far more fish for everyone. These traps are called *collective action problems* (or sometimes *social dilemmas*). Preteens are trapped in a collective action problem when they arrive for their first day of sixth grade and see that some of their classmates have gotten smartphones and are connecting on Instagram and Snapchat, even during class time. That puts pressure on them to get a smartphone and social media, even though all students would be better off if none of them had these things.

Alexis Spence explained to me why she was so desperate to get an Instagram account in sixth grade, despite her parents' prohibition:

> What made it so addictive was that I just wanted to fit in with my peers. I didn't want to miss anything, because if I missed anything, then I was out of the loop, and if I was out of the loop, then kids would laugh at me or make fun of me for not understanding what was going on, and I didn't want to be left out.

Once a few students get smartphones and social media accounts, the other students put pressure on their parents, putting them into a trap as well. It's painful for parents to hear their children say, "*Everyone else* has a smartphone. If you don't get me one, I'll be excluded from *everything.*" (Of course, "everyone" may just mean "some other kids.") Few parents want their preteens to disappear into a phone, but the vision of their child being a social outcast is even more distressing. Many parents therefore give in and buy their child a smartphone at age 11, or younger. As more parents relent, pressure grows on the remaining kids and parents, until the community reaches a stable but unfortunate equilibrium: Everyone really *does* have a smartphone, everyone disappears into their phones, and the play-based childhood is over.

Because technology adoption happens so quickly in the digital world, some of the tech companies find themselves in a collective action problem too. They need to act fast and recruit as many children and teens as possible. Never mind that their own policies and U.S. law require that users be 13 or older; any company that truly verifies the ages of its new users will lose preteens to their competitors, who have no qualms about illegally fishing for underage users.

Parents face collective action problems around childhood independence too. It was easy to send kids out to play back when everyone was doing it, but in a neighborhood where nobody does that, it's hard to be the first one. Parents who let their children walk or play unchaperoned in a public place face the risk that a misguided neighbor will call the police, who may refer the case to Child Protective Services, who'd then investigate them for "neglect" of their children. Each parent decides that it's best to do what every other parent is doing: Keep kids supervised, always, even if that stunts the development of all children.

How do we escape from these traps? Collective action problems require collective responses. There are four main types of collective response, and each can help us to bring about major change:

1. *Voluntary coordination.* Just as parents put additional pressure on the holdouts when they give their 11-year-old a smartphone, parents

can support each other when they stick together. The group Wait Until 8th is a wonderful example of such coordination: Parents sign a pledge when their child is in elementary school that they will not give their child a smartphone until eighth grade. The pledge becomes binding only when 10 families with kids in that school and grade sign the pledge, which guarantees that those children will have others to play with and will not feel that they are the "only ones" excluded. The trap is broken and those 10 families escape together (although only until eighth grade, which is too early because it is still in middle school. I wish they would change their name to Wait Until 9th).

2. *Social norms and moralization.* A community can come to see a personal decision in moral terms and express its revulsion or condemnation, as has happened toward drunk driving (fortunately) or toward a mother who lets her 9-year-old son ride the subway without an adult chaperone (unfortunately).[1] We can reverse the negative moralization of childhood autonomy and come to see 9-year-olds walking around without chaperones as perfectly normal, which it was until very recently.

3. *Technological solutions.* A new product or invention can change the options and incentives for everyone in a community at the same time, such as the introduction of lockable pouches for phones, the development of quick and easy age verification methods, or the introduction of better basic phones, which would reduce the pressure on parents to give their children smartphones and social media before high school.

4. *Laws and rules.* Governments can make laws, such as requiring *all* social media companies to verify the ages of new users, or clarifying neglect laws so that giving independence to a child is not evidence of neglect. Institutions can set policies, such as a school requiring *all* students to keep their phones in phone lockers during the school day.

In the next three chapters, I will lay out a plan by which governments, tech companies, schools, parents, and young people can break out of multiple collective action problems by working together. I called on my friend and collaborator Lenore Skenazy to help me write all three chapters. Lenore is the author of the 2009 book *Free-Range Kids*,[2] which my wife and I read in 2012. It changed the way we raised our children. We gave them independence earlier, which in turn gave them more confidence in themselves, which in turn gave us more confidence in them. I later cofounded an organization with Lenore, along with Peter Gray and Daniel Shuchman, called Let Grow, whose mission is to make it "easy, normal, and legal to give kids the independence they need to grow into capable, confident, and happy adults." You'll notice that some of the sections on rolling back overprotection and increasing play are in a voice different from mine. I thank Lenore for her leadership of the Free-Range Childhood movement and for sharing her wisdom in the coming chapters. Lenore and I highlight a few programs we've developed at Let Grow, but there are many other organizations that share our goals.[3]

A FEW CAVEATS

Before I offer any suggestions, I must include a few notes and acknowledgments.

First, I suggest ideas in the following chapters that I believe will help most families and schools, but every child, family, and school is unique. Most of the psychological principles I draw on are universally applicable, but my suggestions for how to implement them may not be right for you. By all means innovate, improvise, and try to measure the results.

Second, I am surely wrong on some points. I offer advice based on what I wrote in the first eight chapters, which drew on research from many sources. But studies sometimes fail to replicate, social scientists disagree with each other on what studies mean, and new research sometimes points us in new directions. Please consult the online supplement at AnxiousGeneration.com, where I will do my best to correct any errors I have made, and will add additional suggestions. I will also continue to publish essays on my Substack—*After Babel*[4]—where I will present new research and ideas related to this book.

Finally, I want to acknowledge how hard it is to be a parent these days, or a teacher, school administrator, coach, or anyone else who works with children and adolescents. It's even harder to be an adolescent. We're all trying to do our best while struggling with incomplete knowledge about a rapidly changing technological world that is fragmenting our attention and changing our relationships. It's hard for us to understand what is happening, or know what to do about it. But we must do something. We must try new policies and measure the outcomes.

Some of my suggestions are more challenging because they require legislative changes, and, in the United States, political polarization makes it difficult to do anything. But even in the U.S. Congress, protecting children from online harms is one of the few promising areas for bipartisan agreement. If we can understand the nature of collective action problems, we can push for legislation that is targeted at breaking traps and changing incentives. If we act collectively, we can roll back the phone-based childhood and restore, to some degree, a healthier play-based childhood.

Chapter 10

WHAT GOVERNMENTS AND
TECH COMPANIES CAN DO NOW

How do we consume as much of your time and conscious attention as possible?"

That's Sean Parker, the first president of Facebook, in a 2017 interview.[1] He was describing the thought process of the people who created Facebook and the other major social media platforms in the 2000s.

In chapter 2, I quoted another line from this interview, in which Parker explained the "social-validation feedback loop" by which these companies exploit "a vulnerability in human psychology." The apps need to "give you a little dopamine hit every once in a while, because someone liked or commented on a photo or a post or whatever. And that's going to get you to contribute more content, and that's going to get you . . . more likes and comments." He said that he, Mark Zuckerberg, Kevin Systrom (cofounder of Instagram), and others "understood this consciously. And we did it anyway." He also said, "God only knows what it's doing to our children's brains."

Why would anyone treat their customers that way? Because the users are not really the customers for most social media companies.

When platforms offer access to information or services for free, it's usually because the users are the product. Their attention is a precious substance that companies extract and sell to their paying customers—the advertisers. The companies are competing against each other for users' attention, and, like gambling casinos, they'll do anything to hold on to their users even if they harm them in the process. We need to change the incentives so that companies behave differently, as has happened in many other industries. Think of food safety regulations in the Progressive Era, or automotive safety regulations in the 1960s, both of which contributed to the long-running decline in children's mortality rates.[2]

In the first part of this chapter, I describe the ways that many tech companies, particularly social media companies, employ design features that answer Sean Parker's question about how to consume more of people's attention, distracting them from spending needed time in the real world. I then lay out how governments can change laws to incentivize different behavior and design choices, which would make social media less harmful and make it easier for parents to make their own choices about how and when their children enter the virtual world. In the second part of the chapter, I show how governments can change laws and policies that push parents and schools to overprotect in the real world. I also show how governments can make the real world more inviting for children, more supportive of their needs for play, autonomy, and responsibility.

As we'll see, government policies have contributed to the decline of the play-based childhood (especially through the overzealous enforcement of vague state laws about child neglect) and to the rise of the phone-based childhood (especially by setting the age of internet adulthood too low and not enforcing it). New legislation and new enforcement policies would be a tremendous help for parents who are struggling to raise their children in a healthier way.*

---

* I wrote the technology sections of this chapter with assistance from my friend and longtime research collaborator Ravi Iyer, who was a product manager, data scientist, and research manager at Facebook (now Meta) for four years, before leaving to work on technology reform at the University of Southern California Marshall

THE RACE TO THE BOTTOM OF THE BRAIN STEM

Among the keenest analysts of the incentives driving tech companies is Tristan Harris, a former ethicist at Google who, in 2013, created a Power-Point presentation for his fellow Google employees titled "A Call to Minimize Distraction and Respect Users' Attention."[3] Harris noted that the products made by just three companies—Google, Apple, and Facebook—were shaping how most of humanity spent its limited attention, and they were draining it away carelessly or deliberately. The design choices made by tech companies, Harris asserted, had resulted in a global collapse of the amount of attention available for anything beyond screens.

Harris left Google in 2015 and later founded the Center for Humane Technology, an important organization that has been raising the alarm and offering solutions ever since. In 2020, he was invited to testify to a U.S. Senate committee hearing on consumer protection. In his testimony, Harris laid out the incentives companies face in their ferocious competition to extract attention. There are are a number of psychological vulnerabilities that can be abused to capture attention, some of which are grounded in our most basic needs. The companies are stuck in a collective action problem known as a race to the bottom, he said, because if one of them fails to exploit an available psychological weakness, it puts itself at a disadvantage relative to less scrupulous competitors:[4]

> In an attention economy, there's only so much attention and the advertising business model always wants more. So, it becomes a race to the bottom of the brainstem. . . . It starts small. First to get your attention, I add slot machine "pull to refresh" rewards which create little addictions. I remove stopping cues for "infinite scroll" so your mind forgets when to do something else. But then that's not enough. As attention gets more competitive, we have to crawl

School's Neely Center for Ethical Leadership and Decision Making. I also drew on advice from members of two nonprofit tech reform organizations in which I am a participant: Project Liberty, and the Council for Responsible Social Media. I wrote the "real world" sections of this chapter with assistance from Lenore Skenazy.

deeper down the brainstem to your identity and get you addicted to getting attention from other people. By adding the number of followers and likes, technology hacks our social validation and now people are obsessed with the constant feedback they get from others. This helped fuel a mental health crisis for teenagers.[5]

The advertising-driven business model turns users into the product, to be hooked and reeled in. Personalization makes social media companies far more powerful than companies were in pre-digital ad-driven industries such as newspapers and broadcast TV. If we focus on that fact, we can then begin to see where legislation can play a helpful, targeted role, not just regarding social media, but also for video games and pornography sites, which use many of the same attention-grabbing and data-extracting techniques on minors.

For businesses that earn revenue based on displaying ads alongside user-generated content, there are three basic imperatives: (1) get more users, (2) get users to spend more time using the app, and (3) get users to post and engage with more content, which attracts other users to the platform.

One way that companies get more users is by failing to enforce their own rules prohibiting users under 13. In August 2019, I had a video call with Mark Zuckerberg, who, to his credit, was reaching out to a wide variety of people, including critics. I told him that when my children started middle school, they each said that most of the kids in their class (who were 10 or 11 at the start of sixth grade) had Instagram accounts. I asked Zuckerberg what he planned to do about that. He said, "But we don't allow anyone under 13 to open an account." I told him that before our call I had created a fake account for a fictional 13-year-old girl and I encountered no attempt to verify my age claim. He said, "We're working on that." While writing this chapter (in August 2023), I effortlessly created another fake account. There is still no age verification, even though age verification techniques have gotten much better in the last four years,[6] nor is there any disincentive for preteens to lie about their age.

If Instagram were to make a real effort to block or expel underage users, it would lose those users to TikTok and other platforms. Younger users are particularly valuable because the habits they form early often stick with them for life, so companies need younger users to ensure robust future usage of their products. They therefore view the loss of market share among younger users as an existential threat.[7] As a result, companies that make products used by adolescents are trapped in another race to the bottom, a race to get younger and younger users. Documents brought out by the whistleblower Frances Haugen show that Meta has long been trying to study and attract preteens, and has even considered how to reach children as young as 4.[8] (The same race to the bottom occurred with tobacco companies targeting their ads to adolescents, and denying it.)

As for the second imperative, one way that companies get users to spend more time on their apps is by using artificial intelligence to select what to put into a user's feed. Based on the time users spend viewing different kinds of content, AI then serves them more such content.[9] This is why short-form video platforms like TikTok and Instagram Reels are said to be so addictive: Their algorithms are able to quickly detect whatever it is that makes users pause as they scroll, which means they can pick up on unconscious wishes and interests that the user may not even be aware of, leading a minor to be served inappropriate sexual content, for example.[10]

Technology designers long ago learned that reducing friction or effort increases time spent, so features like autoplay and infinite scroll encourage increased consumption of content in an automatic, zombie-like way. When people are asked to identify the platforms on which they spend more time than they want to, the "winners" are social media platforms with these features.[11] Modern video games use different tricks to keep users playing such as free-to-play business models, validation feedback loops, "loot boxes" that are essentially gambling, and never-ending multiplayer games.

To achieve their third objective—incentivizing users to post more content—platforms take advantage of the fact that adolescents are highly sensitive to social status and social rewards. Features like Snapchat "streaks" gamify social interaction by encouraging users to send a picture to their friends every day, in order to not break a publicly visible streak. Snapchat streaks pressure kids to spend more time than they themselves want to spend maintaining network connections, leaving less time for real-world interaction. Another example is setting people's privacy settings to public by default, so that whatever they post becomes content for the largest possible pool of users.

Minors should be protected from products that are designed to addict them. I wish that companies would treat children and adolescents with more care on their own, but given market incentives and business norms, it is likely to take legislation to force them to do so.

WHAT GOVERNMENTS AND TECH COMPANIES CAN DO TO END THE RACE TO THE BOTTOM OF THE BRAINSTEM

There are four main ways that governments and tech companies could improve the virtual world for adolescents.

1. Assert a Duty of Care

In 2013, the British filmmaker Beeban Kidron made a documentary called *InRealLife*, about the lives of teens in the online world. What she learned about the ways tech companies exploit adolescents alarmed her. While she was working on the film, she was awarded a life peerage in the U.K. Parliament's House of Lords, which gave her a new way to act on her concerns. She made online child safety her top priority. After much consultation, she developed a list of design standards that tech companies could adopt that would make time online less harmful to children and adolescents. The list came to be called the Age Appropriate Design Code (AADC), and it was enacted in the U.K. in June 2020.

The code was revolutionary for asserting that companies have some moral and legal responsibility for how they treat minors. They have a duty to design their services in the "best interests" of children, and the code defines children as anyone under the age of 18. For example, it is usually the case that the best interest of the child is served by setting all defaults about privacy to the highest standard, while the best interest of the company is served by making the child's post visible to the widest audience possible. The law therefore requires that the default settings for minors be private; the child must make an active choice to change a setting if she wants her posts to be viewable by strangers. Same thing for geolocation data; the default should be that nobody can find the location of a child from a post or from the use of an app, unless the child elects to make such data public. Another stipulation: Platforms must be transparent and clear about what they are doing, explaining their privacy policies and the nature of parental controls in language (or perhaps videos) easy for children to understand.

While the code applied only to services offered in the U.K., the law has already had two broader effects. First, many of the tech companies decided that it wasn't worth the difficulty to offer different products in different countries, so they made a few of the changes globally. Second, the State of California adopted its own version of the AADC, which was passed into law in 2022, and other states have since passed their own versions.[12] Of course, it makes little sense for individual U.S. states to enact their own laws about something as sprawling and placeless as the internet. It would be far preferable for the U.S. Congress to act, and there is now strong bipartisan support for several important bills, such as the Kids Online Safety Act (KOSA), which includes many ideas from the AADC.[13] But given the longstanding paralysis of U.S. Congress, it has fallen to individual states and governors to try to protect the children in their states from predatory online practices.

Some critics worry that if there is government regulation, it means that the government will be telling people what they can and cannot say on the internet, and the government might well censor one side of the political spectrum or the other. This fear is not unreasonable.[14] But most of the harms platforms are responsible for are not about what *other users are posting* (which is hard for platforms to monitor and control[15]) but about *design decisions* that are 100% within the control of the platforms and that incentivize or amplify harmful experiences.[16] Recent laws such as KOSA are written to focus on design, not content.

Design changes—such as setting privacy preferences to maximum by default—don't give an advantage to either side of the political spectrum. When TikTok limited the ability of teenagers to be contacted by strangers via direct message[17] in response to the U.K. code, or when Facebook pulled back on how advertisers could target underage users with personalized ads,[18] these changes were "viewpoint neutral."[19]

2. Raise the Age of Internet Adulthood to 16

In the late 1990s, as the internet was becoming a part of life, there were no special protections for children online. Companies could collect and sell children's data without the knowledge or consent of their parents. In response, the U.S. Federal Trade Commission recommended that Congress enact legislation requiring websites to obtain parental consent before collecting personal information from children. Representative (now senator) Ed Markey, from Massachusetts, drafted such a bill, and he defined a child as anyone under the age of 16, for data collection purposes. The e-commerce companies of that era objected, and they teamed up with civil liberties groups who were concerned that the new bill would make it harder for teens to find information about birth control, abortion, or other sensitive topics.[20]

In the negotiations over the bill, a compromise was reached that the age would be lowered to 13. That decision had nothing to do with adolescent brain development or maturity; it was just a political compromise. Nonetheless, 13 became the de facto age of "internet adulthood" for the United States, which effectively made it the age of internet adulthood for the world. Anyone who is 13, or at least says they are, can be treated as an adult for the purposes of data acquisition. As Senator Markey later said, "It was too young and I knew it was too young then. It was the best I could do."[21]

In addition to setting the age too low, the bill, known as COPPA (Children's Online Privacy Protection Act), failed to impose any obligation on companies to verify anyone's age. They were only required to avoid collecting data from users when they had direct evidence that the user was under 13. The bill was enacted in 1998, when the internet was a very different place than it is today, and there has been no subsequent action by Congress since then (although several bills are being considered in 2023, including an update of COPPA that would raise the age back to 16).

By specifying 13 as the age of adulthood, COPPA sent a signal to parents that the government thinks 13 is an appropriate age for children to be opening accounts and using these services. It sounds like the "PG-13" by which the Motion Picture Association tells parents that a movie is appropriate for a 13-year-old to see without a parent. But readiness to see a movie is very different from readiness to exercise self-control and make wise choices while being subjected to the addictive attention-extracting techniques used by powerful companies.

What is the right age of internet adulthood? Note that we are *not* talking about the age at which children can browse the web or watch videos on YouTube or TikTok. We're talking only about the age at which a minor can enter into a contract with a company to use the company's products. We're talking about the age at which a child can *open an account* on YouTube or TikTok and begin uploading her own videos and getting her own highly customized feed, while giving her data to the company to use and sell as it says it will do in its terms of service.

Even parents who try hard to keep their children off Instagram often fail, like the mother in Boston whom I quoted in chapter 1, or like Alexis Spence's parents, from chapter 6. When I spoke with the Spences, Alexis's mother described her challenge like this: "I'm fighting AI and I can't beat that. I can't beat a computer that is smarter than me and that is telling her how to outwit me." We can't put the entire burden of policing minimum ages on parents, any more than we would do so when teens try to buy liquor. We expect liquor stores to enforce age limits. We should expect the same from tech companies.

I don't think we should raise the age of internet adulthood all the way to 18. I think Markey's original choice of 16 was the right one for the minimum age at which minors can accept terms of service and give away their data. At 16, adolescents are not adults, but they are more mature and capable than they were at 13. They are also just beyond what may be the most sensitive period for harm from social media (11–13 for girls, 14–15 for boys[22]).

On the other hand, their frontal cortices are still developing, and they are still vulnerable. Social media, video games, porn, and other addictive activities will still harm many of them in a variety of ways. So I'm not saying that the virtual world in its present form, with no guardrails, is *safe* for 16-year-olds. I'm just saying that if we're going to write a minimum age into law and make it an enforceable national standard, then 13 is way too low and 16 is a more reasonable and achievable compromise. It would gain more political and social support than an effort to raise the age to 18. I would just add that 16- and 17-year-olds are still minors, and the protections in any version of an Age Appropriate Design Code should still apply to them. I therefore believe that the U.S. Congress should fix the mistakes it made in 1998 and raise the age of internet adulthood from 13 back to 16, as it was in the original draft of the bill, and then require companies to enforce it.

Now, how can companies do that?

3. Facilitate Age Verification

When people hear the term "age verification," they seem generally to assume it means that users must show a government ID card, such as a driver's license, in order to open an account or access a website. That is one way to do it, and it is the way that Louisiana mandated in a 2023 law. The law required sites whose content is more than one-third pornographic to verify that visitors were over 18, using the state's digital wallet app to present their Louisiana driver's license. Of course, few visitors to a porn site would be willing to give the site their real name, let alone an image of their driver's license. In response, Pornhub simply blocked access to its site from residents who appear to be in Louisiana.

Could social media platforms require an ID of all users to prove that they are old enough to open an account? In theory, yes. States could easily provide identification cards for young people who do not yet have a driver's license. But in practice, platforms get hacked with some regularity, and their databases get sold to thieves or posted on the web, so the threat to privacy would be substantial. Many people would refrain from using valuable services because of it. I am opposed to legally mandating the use of government-issued identity cards for access to parts of the internet run by non-government entities.

Are there ways of verifying age that would still allow people to use a site anonymously? Yes. A second approach to age verification is to have sites farm the job out to another company that simply reports back to the platform: yes or no. Old enough, or not old enough.[23] If the age verification company was hacked, all that the world would learn is that the people in their database once had their age verified, not that they had used Pornhub or any other site.

Companies have developed methods such as these:

- Using a network of people to vouch for each other (such that those who lie lose the privilege of vouching).

- Issuing a blockchain token to anyone who is verified once, by a reliable method. The token then serves like a driver's license to prove that the bearer is above a certain age, but it carries no personal information about the bearer, so a data hack would reveal nothing.

- Using biometrics to establish identity. Clear, a company known for rapid identity verification in airports, is now used as a quick way for its clients—who previously verified their age—to prove that they are old enough to buy alcohol at stadium events.

So many companies now offer different methods of age verification that they have their own trade association.[24] The quality, reliability, and security of these methods are sure to increase over time. I hope that companies that want to enforce a minimum age will begin to offer a *menu* of options from which the user can pick.[25] Some of the methods would take only a few seconds. Laws such as the one in Louisiana would create far smaller privacy concerns if they allowed companies to offer a menu of reliable options, rather than mandating the use of a government-issued ID.

There is not, at present, any perfect method of implementing a universal age check. There is no method that could be applied to everyone who comes to a site in a way that is perfectly reliable and raises no privacy or civil liberties objections.[26] But if we drop the need for a *universal* solution and restrict our focus to helping parents who want the internet to have age gates that apply *to their children*, then a third approach becomes possible: Parents should have a way of marking their child's phones, tablets, and laptops as devices belonging to a minor. That mark, which could be written either into the hardware or the software, would act like a sign that tells companies with age restrictions, "This person is underage; do not admit without parental consent."

A simple way to do this would be for Apple, Google, and Microsoft—who create the operating systems that run nearly all of our devices—to add a feature to their existing parental controls. In Apple's iOS, for example, parents already set up family accounts and put in the correct birth dates for their children when they give them their first iPhones. The parent already gets to choose whether to allow the child to download only age-appropriate apps, movies, and books from Apple's *own* services. Why not just expand that ability so that a parent's choice is respected by *all* platforms for which age restrictions are appropriate, or required by law? (Parents already have the ability to block access to specific websites, but that puts the onus on parents to know what sites and categories of sites they want to block, which parents can't know unless they monitor their children's online activity closely and monitor online sites and trends.[27])

Apple, Google, and Microsoft could create a feature, let's call it *age check*, which would be set to "on" by default whenever a parent creates an account for a child under the age of 18. The parent can choose to turn age check off, but if on is the default, then it would be very widely used (unlike many features in current parental controls, which many parents don't know how to turn on). If age check is left on, then when anyone uses that phone or computer to try to open an account or log in to an account, the site can simply verify by communicating with the device to answer two questions: 1) Is age check on? If so, then 2) Does the person meet our minimum age? (For example, 16 to open or access a social media account, 18 to access pornography.)

This kind of device-based verification offers a way that parents, tech companies, and platforms can share responsibility for age verification. Such a system would have helped Alexis Spence's parents to keep their 10-year-old daughter off the social media platforms that took over her life. It would also have reduced the peer pressure on Alexis because few of her classmates would have been on Instagram. It would also allow sites to age gate specific features, such as the ability to upload videos or to be contacted by strangers. Note that with device-based verification, *nobody else is inconvenienced*. Adults who visit a site that uses age check don't have to do anything or show anything, so the internet is unchanged for them, and there is no privacy threat whatsoever. Parents who want their children to open social media accounts or visit pornography websites can simply turn age check off.

4. Encourage Phone-Free Schools

In the next chapter—on what schools can do—I'll make the case that all schools, from elementary through high school, should go phone-free to improve not only mental health but academic outcomes as well. Governments at all levels, from local to federal, could support this transition by allocating funds to pay the small cost of buying phone lockers or lockable pouches for any school that wants to keep phones out of students' pockets and hands during the school day. Departments of education at the state and federal levels could support research on the effects of phone-free schools, to verify whether they are beneficial for student mental health and academic performance.

WHAT GOVERNMENTS CAN DO TO INCENTIVIZE MORE (AND BETTER) REAL-WORLD EXPERIENCE

During the summer of 2014, when the South Carolina single mom Debra Harrell worked her shifts at McDonald's, she brought along her daughter, who was on vacation from school. Regina, age 9, spent the time playing on a laptop. But when the laptop was stolen from their home, Regina begged her mom to let her play at the neighborhood's popular sprinkler park instead. She'd be surrounded by friends and many of their parents. It felt safe. It felt like summer. Debra said yes.

But on Regina's third day of fun in the sun, a woman at the park asked her where her mom was. When she said, "Working," the woman called 911. The police charged Debra with child abandonment—which carries up to a 10-year sentence—and threw her in jail. Regina was taken away from her mom for 17 days.[28]

This case and many others like it frighten parents into over-supervising their children. Governments are literally criminalizing the play-based childhoods that were the norm before the 1990s.

1. Stop Punishing Parents for Giving Children Real-World Freedom

Debra's experience and other stories of parents investigated for things like letting children play outside[29] or get themselves home from the park[30] led Let Grow to start a movement for "Reasonable Childhood Independence" laws. Currently, neglect laws in most states are vague, saying things like "Parent must provide proper supervision." Yes, of course they should, but people have wildly different ideas of what that entails. Just because some passerby wouldn't let *her* nine-year-old play outside doesn't mean the state should be able to investigate anyone who does.

A study in *Social Policy Report* found that the way current U.S. laws are written and interpreted has little relationship to the ages at which children develop abilities.[31] In societies around the world, children were traditionally thought to become much more capable and responsible around the ages of 6 or 7, when they were routinely given responsibilities such as caring for younger children and animals. Yet in some U.S. states, such as Connecticut, the law said a child should never be left alone in public before the age of 12, meaning that 11-year-olds needed babysitters. Indeed, a Connecticut mom was arrested for letting her 11-year-old wait in the car while she ran into the store.[32] This, despite the fact that the Red Cross begins training babysitters at the age of 11, which is the age at which my sisters and I began to babysit for neighbors. Let Grow lobbied successfully for Connecticut to change its criminal endangerment law in 2023. But other states' neglect laws remain ambiguous, allowing the authorities wide discretion to intervene.

The *Social Policy Report* essay notes, "Parents who fail to provide their children opportunities for physical and cognitive stimulation through independent activities are potentially 'neglecting' their children in those dimensions." So a lack of adult supervision should not be the touchstone for a neglect finding. In fact, maybe the state is engaging in neglect when it mandates overprotection.

Reasonable Childhood Independence laws clarify the meaning of neglect: Neglect is when a parent blatantly, willfully, or recklessly disregards a danger to a child so apparent that no reasonable person would allow the child to engage in that activity. In other words, it is not neglect when you simply take your eyes off your children. This clarification protects parents who give their kids more independence for its own sake, as well as those who do so out of economic necessity, like Debra Harrell.

In 2018, Utah became the first U.S. state to pass such a law. Since then, Texas, Oklahoma, Colorado, Illinois, Virginia, Connecticut, and Montana have too. These bills have usually had bipartisan sponsors, and they often passed unanimously. They appeal to people across the political spectrum because no one wants the government meddling in family life if there is not a compelling reason.

The government's job is to protect children from actual abuse, not from the everyday activities of childhood. States must revise their supervisory neglect laws. They must cease and desist all enforcement action against parents whose only offense is that they chose to give their children reasonable independence, appropriate for their age. Ask your state legislators (or equivalent in other countries) to introduce a Reasonable Childhood Independence law.[33]

2. Encourage More Play in Schools

In the next chapter, I'll make the case that schools in the United States are starving children of playtime in order to make ever more room for academic training and test preparation, which backfires because play-deprived kids become anxious and unfocused. Ultimately, they learn less. Governors' offices and state education departments should take seriously the research on the benefits of free play in general and recess in particular.[34] Then they should mandate that schools give a lot more of it, including play opportunities before and after school, especially in elementary and middle school.[35]

3. Design and Zone Public Space with Children in Mind

If we want children to meet each other face-to-face and interact with the real world—not just screens—the world and its inhabitants have to be accessible to them. A world designed for automobiles is often not one that children find accessible. Cities and towns can do more to be sure that they have good sidewalks, crosswalks, and traffic lights. They can install traffic calming measures, and they can change their zoning to allow more mixed-use development. When commercial, recreational, and residential establishments are more mashed up together, there is more activity on the street and more places that children can get to on foot or by bike. But when the only way for a kid to get to a shop, park, or friend's house is by "parent taxi," more kids will end up at home on a screen. One study found that kids who can get to a playground by bike or foot are six times more likely to visit it than kids who need someone to drive them.[36] So scatter playgrounds throughout a neighborhood, and consider having a few of them be adventure playgrounds (see next chapter).

One innovative and inexpensive way that European cities are helping kids (and parents) to be more sociable is by blocking off the street in front of a school for an hour before and after school.[37] On these tempo-rarily car-free School Streets, parents mingle and kids play, even as con-gestion, pollution, and road danger go down. Cities can make this happen by easing the street closure permitting process. In our era of declining community and rising loneliness, cities and towns should make it easy for local residents to block off streets for block parties and other social reasons too, including Play Streets (streets closed to traffic, part time, so kids can play with each other, like old times).[38]

When considering transit, zoning laws, permits, and new construc-tion, remember that kids are human beings. They want to be where the action is. Easily accessible mixed-use spaces where everyone, young and old, can hang out, see, be seen, do some playing, shopping, eating, flirt-ing, and, when tired, bench sitting make everyone more engaged with the world beyond the screen.

4. More Vocational Education, Apprenticeships, and Youth Development Programs

The educational system in the United States has become ever more fo-cused on academic training that leads to a college education. There has been a corresponding decline in course offerings and student participa-tion in what is known as career and technical education, or CTE. These are courses with a lot of hands-on experience in areas such as shop, auto mechanics, agriculture, and business. Richard Reeves says the research is strongest on the benefits of sending boys to specialized high schools devoted to CTE. Boys in such schools saw big gains in their graduation rates and later earnings, compared with similar boys who attended tra-ditional high schools, while girls did not show these particular benefits.[39] These findings are further evidence that standard schools are failing to engage many boys, leading to enormous wasted potential.

Apprenticeships have also been shown to be effective for helping young people make the transition from high school to paid employment. In a labor market in which people move around frequently, companies have little incentive to take on untrained young people, invest in them, and then have them move elsewhere. Government-supported programs that subsidize pay for a period of time make it less expensive for companies to train young people, thereby increasing their value to the company or any future employer.[40]

Governments can also support gap year and "year of service" programs, particularly among young people who do not have clear college prospects. Programs like AmeriCorps help young people learn new skills while helping local communities. Wilderness experience programs have also been shown to confer benefits to adolescents;[41] they offer direct training in antifragility while immersing young people in natural beauty. Such programs are usually run by nonprofit or for-profit companies, but the State of Connecticut has been running a tuition-free program since 1974 for adolescents across the state.[42]

GOVERNMENTS HAVE THE POWER AND OFTEN THE RESPONSIBILITY TO address collective action problems. Poorly crafted and erratically enforced laws have exacerbated some of these problems. Governments can set standards that change the behavior of companies. They can set age limits that shut off competition for underage users. They can make it easier for parents and schools to grant more freedom to children and adolescents, as I'll discuss in the next two chapters. When governments, tech companies, schools, and parents work in complementary ways, they can collectively solve hard problems, including improving the mental health of young people.

IN SUM

- Governments at all levels need to change policies that are harming adolescent mental health and support policies that would improve it. In the United States, governments at the state and local level are partly responsible for the overprotection of children in the real world (via vast overreach of vague neglect laws), and the federal government is partly responsible for the underprotection of children in the virtual world (by passing an ineffective law in 1998 and failing to update it as the dangers of life online became more apparent).

- To correct underprotection online, national and federal governments should enact laws of the sort first passed in the U.K., which require companies to treat minors differently than adults, with an extra duty of care. National governments should also raise the age of internet adulthood to 16.

- Tech companies can be a major part of the solution by developing better age verification features, and by adding features that allow parents to designate their children's phones and computers as ones that should not be served by sites with minimum ages until they are old enough. Such a feature would help to dissolve multiple collective action problems for parents, kids, and platforms.

- To correct overprotection in the real world, state and local governments should narrow neglect laws and give parents confidence that they can give their children some unsupervised time without risking arrest or state intervention in their family life.

- State and local governments should also encourage more free play and recess in schools. They should consider the needs of children in zoning and permitting, and they should invest in more vocational education and other programs that have been shown to help adolescents, especially boys, make the transition to adulthood.

Chapter 11

WHAT SCHOOLS CAN DO NOW

n April 2023, *The Washington Post* ran a story with the headline "One School's Solution to the Mental Health Crisis: Try Everything."[1] It was about a K–12 school in rural Ohio whose leaders had brought in more therapists and purchased a new social-emotional learning curriculum offering formal instruction in "qualities like empathy and trust, and skills like relationship-building and decision-making." The school encouraged children as young as kindergarten to sing about their emotions in music class. They brought in horses for children to pet and groom in after-school care, courtesy of an organization that promotes trauma-sensitive experiential learning.

There is a Polynesian expression: "Standing on a whale, fishing for minnows." Sometimes it is better to do a big thing rather than many small things, and sometimes the big thing is unnoticed but right underfoot. To address the widespread anxiety in this generation, there are two whales—two big things that schools could do using mostly resources they already have. These are phone-free schools and more free play. If they are done together, I believe they would be more effective than all of the other measures schools are now taking, combined, to improve the mental health of their students.

PHONE-FREE SCHOOLS

Mountain Middle School in Durango, Colorado, went phone-free back in 2012, at the start of the mental health crisis. The county around the school had among the highest teen suicide rates in Colorado when Shane Voss took over as head of school. Students were suffering from rampant cyberbullying, sleep deprivation, and constant social comparison.[2]

Voss implemented a cell phone ban. For the entire school day, phones had to stay in backpacks, not in pockets or hands. There were clear policies and real consequences if phones were found out of the backpack during school hours.[3] The effects were transformative. Students no longer sat silently next to each other, scrolling while waiting for homeroom or class to start. They talked to each other or the teacher. Voss says that when he walks into a school without a phone ban, "It's kind of like the zombie apocalypse, and you have all these kids in the hallways not talking to each other. It's just a very different vibe."

The school's academic performance improved, and after a few years it attained Colorado's highest performance rating. An eighth grader named Henry explained the effect of the phone ban. He said that for the first half hour of the school day his phone is still in the back of his mind, "but once class starts, then it's just kinda out the window and I'm not really thinking about it. So it's not a big distraction for me during school." In other words, the phone ban ameliorates three of the four foundational harms of the phone-based childhood: attention fragmentation, social deprivation, and addiction. It reduces social comparison and the pull into the virtual world. It generates communion and community.

Naturally. Smartphones and their apps are such powerful attention magnets that half of all teens say they are online "almost constantly." Can anyone doubt that a school full of students using or thinking about their phones almost all the time—texting each other, scrolling through social media, and playing mobile games during class and lunchtime—is going to be a school with less learning, more drama, and a weaker sense of community and belonging?

Most public schools in the United States *say* that they ban phones; 77% said so in a 2020 survey.[4] But that usually just means that the school forbids phone use *during class time*, so students must hide their phones in their laps or behind a book in order to use them. Even if such a ban were perfectly enforced by hypervigilant teachers patrolling each row of the classroom, it would mean that the moment class ended, most students would pull out their phones, check their texts and feeds, and ignore the students next to them. When students are allowed to keep their phones in their pockets, phone policing becomes a full-time job, and it is the last thing that teachers need added to their workload. Many of them eventually give up and tolerate open use.[5] As one middle school teacher wrote to me, "Give teachers a chance. Ban smartphones."

A phone "ban" limited to class time is nearly useless. This is why *schools should go phone-free for the entirety of the school day.* When students arrive, they put their phone into a dedicated phone locker or into a lockable phone pouch. At the end of the day, they retrieve their phones from the locker, or they access a device that unlocks the pouch. (Some parents object that they need to be able to reach their children immediately in case of an emergency, such as a school shooting. As a parent I understand this desire. But a school in which most students are calling or texting their parents during an emergency is likely to be less safe than a school in which only the adults have phones and the students are listening to the adults and paying attention to their surroundings.[6])

The evidence that phones in pockets interfere with learning is now so clear that in August 2023, UNESCO (the United Nations Educational, Scientific, and Cultural Organization) issued a report that addressed the adverse effects that digital technologies, and phones in particular, are having on education around the world.[7] The report acknowledged benefits of the internet for online education and educating some hard-to-reach populations, but noted that there is surprisingly little evidence that digital technologies enhance learning in the typical classroom. The report also noted that mobile phone use was associated with reduced educational performance and increased classroom disruption.[8] So going phone-free is a crucial first step. Each school would still need to consider the effects of laptops, Chromebooks, tablets, and other devices through which students can text each other, access the internet, and be pulled away into digital distractions. The value of phone-free and even screen-free education can be seen in the choices that many *tech executives* make about the schools to which they send their own children, such as the Waldorf School of the Peninsula, where all digital devices—phones, laptops, tablets—are prohibited. This is in stark contrast with many public schools that are advancing 1:1 technology programs, trying to give every child their own device.[9] Waldorf is probably right.

Additional evidence that phones may be interfering with education in the United States can be found in the 2023 National Assessment of Educational Progress (otherwise known as the nation's report card), which showed substantial drops in test scores during the COVID era, erasing many years of gains. However, if you look closely at the data, it becomes clear that the decline in test scores began earlier.[10] Scores had been rising pretty consistently from the 1970s until 2012, and then they reversed. COVID restrictions and remote schooling added to the decline, especially in math, but the drop between 2012 and the beginning of COVID was substantial. The reversal coincided with the moment teens traded in their basic phones for smartphones, leading to a big increase in attention fragmentation throughout the school day. But it wasn't Kurt Vonnegut's egalitarian dystopia, where the top students had to wear an earpiece that disrupted their thoughts. Instead, it was the students in the *lower quarter* whose scores dropped the most between 2012 and 2020. These students are disproportionately from lower-income households, with Black and Latino students overrepresented.

Studies show that lower-income, Black, and Latino children put in more screen time and have less supervision of their electronic lives, on average, than children from wealthy families and white families. (Across the board, children in single-parent households have more unsupervised screentime.[11]) This suggests that smartphones are *exacerbating educational inequality* by both social class and race. The "digital divide" is no longer that poor kids and racial minorities have less *access* to the internet, as was feared in the early 2000s; it is now that they have less *protection* from it.

But smartphones don't just damage learning. They also damage social relationships. In chapter 1, I showed that students across the globe suddenly began to disagree more often, after 2012, with statements like "I feel like I belong at school." Because adolescents today are starving for community and communion, phone-free schools are likely to bring about a rapid improvement in school socializing and mental health.[12]

Of course, the internet itself is a boon to education; just think about the profound global good done by a platform like Khan Academy. Look at how Khan Academy is now using AI to give every student their own personal tutor, and every teacher their own assistant.[13] Furthermore, students need the internet to do research, and teachers need the internet for many innovative lessons, demonstrations, and videos. Schools should help students learn to code and to use technology that expands their abilities, from statistical software through graphic design and even ChatGPT.

So I would never say that we need internet-free schools or students. It's the personal devices that students carry with them throughout the school day that have the worst cost-benefit ratio. Students' phones are loaded with apps designed to catch the attention of young people, pinging them with notifications calling them out of class and into their virtual worlds. That's what is most disruptive to learning and relationships. Any school whose leaders say that they care about fostering belonging, community, or mental health, but that hasn't gone phone-free, is standing on a whale, fishing for minnows.

PLAY-FULL SCHOOLS

Kevin Stinehart, a fourth-grade teacher at the Central Academy of the Arts, an elementary school in rural South Carolina, realized he was having the same conversation over and over with teachers and parents. Students were struggling, and many seemed to have little resilience, perseverance, or ability to work with others. The adults were all talking about the students' fragility, but none had any idea what to do about it. Kevin was stumped too, until he attended a conference at nearby Clemson University on the benefits of something pretty basic: free play. With his school's blessing and help from Let Grow, Kevin started to incorporate more free play into students' lives by making three changes:

1. Longer recess with little adult intervention.

2. Opening the school playground for half an hour before school starts, to give students time to play before class.

3. Offering a "Play Club." Anywhere from one to five days a week, a school stays open for mixed-age, "loose parts" free play (featuring things like balls, chalk, jump ropes), usually on the playground, or in the gym in bad weather. (But if the school can keep other rooms open, like an art room, great!) From 2:30 to 4:30 p.m.—your schedule may vary—instead of going home (often to a device or to an adult-led activity), children spend time together playing. It's a no-phone zone! The kids are given nearly complete autonomy. There are only two rules: They can't deliberately hurt anyone; and they can't leave without telling the person in charge. This adult doesn't organize any games or solve any spats. Like a lifeguard, adults intervene only in the case of an emergency. (Let Grow provides a free Play Club implementation guide on its website.)

In the very first semester he made these changes, Kevin started noticing a shift in his students:

Our students are happier, kinder, have fewer behavior problems, have made more friends, feel more in control of their day and their life in general, and in some cases have dramatically changed course from bullying behaviors and frequent office referrals to no bullying behaviors and no office referrals.[14]

The next semester, he offered Play Club twice a week because "the benefits we were seeing were too huge to ignore." How huge? Compared with the previous year, truancy cases went from a total of 54 down to 30, and school bus violation incidents dropped from 85 to 31. "In any given school year we used to have around 225 office referrals," Kevin reported. "But now that we've added so much play we only have around 45."

Kevin believes that the Play Club caused these changes for the following reasons:

> Unstructured free play addresses—head-on—making friends, learning empathy, learning emotional regulation, learning interpersonal skills, and greatly empowers students by helping them find a healthy place in their school community—all while teaching them life's most important skills like creativity, innovation, critical thinking, collaboration, communication, self-direction, perseverance, and social skills.

The teachers themselves saw such a big change that 13 ended up volunteering to supervise the Play Club. So did the principal and assistant principal.

Note how free play achieves many of the social-emotional learning goals sought by the Ohio school that had "tried everything." At the Ohio school, social emotional learning is taught by adults as yet one more structured curriculum. Free play at the Central Academy of the Arts, in contrast, brought rapid learning because it is nature's way of teaching these same skills as a side effect of kids doing what they most want to do: Play with each other.

THOSE ARE THE TWO WHALES: GOING PHONE-FREE AND GIVING A LOT more unstructured free play. A school that is phone-free and play-full is investing in prevention. It is reducing overprotection in the real world, which helps kids to cultivate antifragility. At the same time, it is loosening the grip of the virtual world, thereby fostering better learning and relationships in the real world. A school that does neither is likely to struggle with high levels of student anxiety, and will need to spend large amounts of money to treat students' growing distress.

I'll offer just a few additional actions schools can take that would complement going phone-free and becoming play-full.[15]

THE LET GROW PROJECT

Many American children, even in middle school, have never been allowed to walk beyond their block or drift a few aisles away from their parents in a large store. Lenore has met seventh graders—kids 12 and 13—who have never been allowed to cut their own meat because sharp knives are dangerous.

That's why, in addition to starting a Play Club, lengthening recess, and opening the playground before school starts, Lenore and I recommend that schools assign the Let Grow Project.[16] It is a homework assignment that tells students from kindergarten through middle school, "Go home and do something you've never done on your own before. Walk the dog. Make a meal. Run an errand." Students confer with their parents, and both generations agree on what the project will be.

When the child succeeds—which they almost always do, eventually—relationships and identities begin to change. The parents see their children as more competent, and so do the kids themselves. By gently pushing parents to give their kids a little more independence (and thus, responsibility), the project addresses a specific problem. Many parents have no idea when they can start letting kids do things on their own, so they just don't. In earlier eras, kids as young as 5 were walking to school. Their crossing guards were 10-year-olds with all the traffic-stopping power of an orange sash. But those independence milestones gradually disappeared under a mountain of media-fueled fear.

We shouldn't blame parents for "helicoptering." We should blame—and change—a culture that tells parents that they *must* helicopter. Some schools won't let kids get off the school bus unless there's an adult waiting there to walk them home.[17] Some libraries won't allow kids under age 10 to wander beyond their parents' sight lines.[18] And some parents have been arrested simply for letting their kids play outside or walk to the store. When parents can't take their eyes off their kids, and kids can't do a single thing on their own, the result is a double helix of anxiety and doubt. Many kids are afraid to try something new, and their parents don't have confidence that they can do something new, all of which leads to more overprotection, which leads to more anxiety.

That's what Lenore heard when she visited a seventh-grade health class in Suffolk County, New York. Veteran teacher Jodi Maurici told her, "Their parents have just made them so scared of everything." The students in this class were sweet and open but feared that anything they might try on their own could end up a disaster. Many said they'd been afraid to cook because they didn't want to burn the food (or house). A few said they'd been afraid to walk the dog because what if it ran into the street? Some of the kids had been afraid to talk to a waitress because they might "mess up"—a phrase they used over and over. Everyday life was a minefield of potential failure and humiliation. (Kind of like social media.) That's why Jodi assigned the Let Grow Project.

In fact, Jodi was so worried about her students' anxiety levels she had each of them do 20 Let Grow Projects over the course of the year. She gave them a long list of things to choose from: Walk to town, do the laundry, ride a bus . . . and of course they could add their own. As the year was drawing to a close, Jodi had seen such a drop in her students' anxiety levels that she invited Lenore to spend an afternoon talking to the kids about their projects.

One girl told Lenore she went to the park with friends for the first time without a parent. "It was *so much fun!*" A boy who cooked a four-course dinner, including baking a pie, felt incredibly accomplished. A girl who'd never done any sports tried out for the swim team and made it. Kids were going out for pizza, biking to the store, babysitting, and feeling something completely new. It wasn't just a new sense of confidence. It was a new sense of who they were, which one girl explained without realizing it. Her favorite project, out of the 20 done that year, was the time she was allowed to stay home one morning without her parents and get her 5-year-old sister ready for school.

Once she got the little girl dressed and fed and put her on the bus, the seventh grader said, "I felt so grown-up!" But it wasn't just that. "It seems small. But in the moment, when I saw her get on the bus and it drove away, I felt really important to her, important to someone." That's what was so new to her. At last, instead of feeling needy, she was needed.

When we give trust to kids, they soar. Trusting our kids to start venturing out into the world may be the most transformative thing adults can do. But it is difficult for most parents to do this on their own. If your daughter goes to the park and there are no other kids there, she'll come right home. If your son is the only 8-year-old anyone in your town ever sees walking without a chaperone, someone might call the police.

Re-normalizing childhood independence requires collective action, and collective action is most easily facilitated by local schools. When an entire class, school, or school district encourages parents to loosen the reins, the culture in that town or county shifts. Parents don't feel guilty or weird about letting go. Hey, it's homework, and all the other parents are doing it too. Pretty soon, you've got kids trick-or-treating on their own again, and going to the store, and getting themselves to school.

Our kids can do so much more than we let them. Our culture of fear has kept this truth from us. They are like racehorses stuck in the stable. It's time to let them out.

BETTER RECESS AND PLAYGROUNDS

There are three big ways to improve recess: Give kids more of it, on better playgrounds, with fewer rules.

We should all be aghast that the average American elementary school student gets only 27 minutes of recess a day.[19] In maximum-security federal prisons in the United States, inmates are guaranteed two hours of outdoor time per day. When a filmmaker asked some prisoners how they'd feel if their yard time was reduced to one hour, they were very negative. "I think that's going to build more anger," said one. "That would be torture," said another. When they were informed that most children around the world get less than an hour a day of outdoor playtime, they were shocked.[20]

Recess in America—and children's unstructured time outside school—has been shrinking ever since the publication of a landmark 1983 report titled *A Nation at Risk*. The report warned that American kids were falling behind those of other nations in test scores and academic proficiency.[21] It recommended increasing rigor by spending more time on academic subjects and considerably lengthening the school year. Schools responded. Time allotted for recess, gym, art, and music classes all decreased, to make way for more math, science, and English.

While the *Nation at Risk* report *did not* call for a single-minded focus on test scores, in practice that is what happened. Raising test scores quickly became a national obsession as new reform efforts penalized or rewarded schools based on test performance. The pressure on schools to deliver rising scores increased again after the 2001 federal No Child Left Behind Act was passed, and more recently the Common Core State Standards.[22] (The pressure was so intense that some school districts met their targets by simply falsifying their students' test scores.[23]) Children's playtime was the easiest activity to sacrifice to make room for more drills and test prep. School years lengthened (cutting into summer vacation), homework levels increased (and got pushed down to lower grades), and recess got shortened or cut entirely.

As a professor, I'm certainly in favor of reforms that increase academic performance, but the preoccupation with test scores caused the educational system to violate much of what we know about child development, the benefits of free play, and the value of time outdoors. The American Academy of Pediatrics issued a report in 2013 titled "The Crucial Role of Recess in School." After describing the many benefits of free play for social and cognitive development, it said, "Ironically, minimizing or eliminating recess may be counterproductive to academic achievement, as a growing body of evidence suggests that recess promotes not only physical health and social development but also cognitive performance."[24] These benefits may be particularly large for boys,[25] which suggests one more reason why boys have increasingly disengaged from school since the 1970s.

The first thing schools can do to improve recess is to give students more of it. Generous recess should extend through middle school, and some recess should be given even in high school (according to the U.S. Centers for Disease Control[26]). The American Academy of Pediatrics also recommends that schools *not* use revoking recess as a punishment for bad behavior, in part because it is precisely the kids with behavioral problems who need recess most. Its report also recommends giving recess *before* lunch, rather than the common practice of combining lunch and recess as a single short period in which students wolf down their food in order to maximize their few precious minutes of free play.

The second way to improve recess is to improve the playground. A typical playground in the United States, especially in cities, is just asphalt with a few metal or plastic play structures that were designed for durability and safety. Often there is also a grassy area for sports. But Europeans have led the world in designing what are known as *adventure playgrounds*, which are designed for imaginative play. One type is called a junk playground because it is filled with miscellaneous things—building materials, ropes, and other "loose parts," often along with tools, which attract children like magnets.

New York City is blessed with one such playground on Governors Island, the best playground my children have ever experienced.[27] Signs encircling the playground (see figure 11.1) tell parents to refrain from interfering. As a parent, I know that is hard. When anyone sees a kid struggling, they want to jump in to help. It's normal. It's the natural outcome of being present and seeing a child who's frustrated or taking a small risk or behaving badly. That is why it is so important that we carve out some time when kids are *not* with a parent, teacher, or coach. That's pretty much the only time they will be forced to function on their own and realize how much they are capable of.

At the adventure playground, children work together to build towers and forts, deeply engrossed in joint activities. On one visit, I saw a boy hit his thumb while hammering nails, but he didn't run to an adult. He just shook his hand for a few second and returned to hammering. (There are adults on-site, monitoring for any serious safety risks.)

While schools need not turn their playgrounds into junkyards, they can add loose parts to the mix. Not necessarily hammers and saws, but things like tires, buckets, and loose boards. Rusty Keeler, author of *Adventures in Risky Play*, also recommends things like hay bales and sandbags. These are so big and heavy that dragging them around "sneaks in upper body strength," he says.[28] And because one kid can't move a bale alone, kids end up working together, seamlessly building social development and collaboration into recess. The key thing to understand about "loose parts" playgrounds is that kids have control over their environment. They have agency. Playgrounds with fixed structures can hold kids'

Figure 11.1. The junkyard playground on Governors Island, New York City, designed and run by play:groundNYC.[29]

Figure 11.2. A state-of-the-art nature playground just before its opening in 2023 at Colene Hoose Elementary School in Normal, Illinois.[30]

attention only so long. But loose parts keep kids' attention for hours, allowing them to build not only forts and castles but also focus, compromise, teamwork, and creativity.

A second category of adventure playground is a nature playground, as in figure 11.2, which uses natural materials, especially wood, stone, and water, to create environments that activate the "biophilia" (that is, "love of life") response that I described in chapter 8.

Human childhood evolved in savannas and forests, alongside streams and lakes. When you put children into natural settings, they instinctively explore and spontaneously invent games. Abundant research shows that time in natural settings benefits children's social, cognitive, and emotional development,[31] and these benefits matter even more as young people are increasingly ensconced in the virtual world and as their anxiety levels continue to rise. One review of studies on the effects of nature playgrounds concluded:

> Providing young people with opportunities to connect with nature, particularly in educational settings, can be conducive to enhanced cognitive functioning. Schools are well placed to provide

much needed "green" educational settings and experiences to assist with relieving cognitive overload and stress and to optimize wellbeing and learning.[32]

The third way to improve mental health by improving recess is to reduce rules and increase trust. Essentially, schools should do the opposite of the school in Berkeley, California, that I discussed in chapter 3. That's the school that specified exactly how children should play tag, four square, and touch football, including the rule that students must not attempt to play touch football without an adult referee.

To see what the opposite of the Berkeley school looks like, consider the "No Rules Recess" at Swanson Primary School in New Zealand.[33] Before "No Rules Recess," students had been forbidden to climb trees, ride bikes, or do anything with any risk. But then the school took part in a study in which researchers asked eight schools to reduce rules and increase opportunities for "risk and challenge" during recess, while eight other schools were asked to make no changes to their recess policies. Swanson was in the freedom group, and Principal Bruce McLachlan decided to go all the way: He scrapped *all* rules and let kids make their own.

The result? More chaos, more activity, more pushing and shoving on the playground, and also more happiness and more physical safety. Rates of injury, vandalism, and bullying all declined,[34] just as Mariana Brussoni and other play researchers had been saying would happen.[35] Kids will take on responsibility for their safety when they *are actually responsible* for their safety, rather than relying on the adult guardians hovering over them.[36]

Could elementary schools in the United States follow Swanson's example? Right now, few could. At many schools, the threats of lawsuits and parental protests are too great. The fear that this will take away from test prep is too high. That's why this is a collective action problem: Students would be healthier, happier, and smarter overall, with lower rates of injury and anxiety, if schools could loosen the reins and let children play in a more natural way. But we can't get there unless schools, parents, and governments can find a way to work together.

RE-ENGAGE BOYS

Boys' and young men's success has been declining, on some measures, since the 1970s. I suggested in chapter 7 that the decline was caused by their gradual disengagement from the real world (due to a variety of structural forces), while they were simultaneously enticed into the virtual world by ever-improving technologies that appeal to boys' desires. As Richard Reeves has shown, boys are lagging behind in academic achievement, graduation rates, college degrees, and almost every measure of educational outcomes. Schools are just not working for a growing number of boys.

Reeves offers a number of policy reforms that would help reverse the trends for boys, including more vocational training and CTE, as I discussed in the previous chapter. In addition, Reeves urges schools to hire more male teachers. He notes that the male share of K–12 teachers in the United States is just 24%, down from 33% in the early 1980s. In elementary schools, just 11% of teachers are male. He notes two ways that this hastens the disengagement of boys. First, there is solid evidence that boys do better academically when they have a male teacher, especially in English classes.[37] This may be due to role model effects, because boys often have few male role models in school. As one progressive education analyst wrote, "Having both male and female teachers is likely good for students for many of the same reasons that they benefit from a racially and ethnically diverse teacher workforce."[38] Without positive male role models in their lives, many boys turn online for guidance, where they are easily led down rabbit holes into online communities that may radicalize their thinking.

A second way that the gender imbalance in education harms boys is that it teaches them—most strongly in elementary school—that the education and caring professions are female. This makes boys less likely to have an interest in these professions. Yet as Reeves points out, it is precisely in these professions that job growth has been strong for decades, and will continue to grow, while the more male-coded jobs requiring physical strength will continue to decline in number. Reeves thinks that schools can and should steer boys into the "HEAL" professions, which stands for health, education, administration, and literacy.[39] But as long as boys see few men in front of the classroom or working as administrators in the school, they're likely to have less interest in such jobs.

THE EDUCATION EXPERIMENT WE MOST NEED

In May 2019, I was invited to give a lecture at my old high school in the suburbs of New York City. Before the talk, I met with the principal and his top administrators. I heard that the school, like most high schools in America, was struggling with a large and recent increase in mental illness among its students. The primary diagnoses were depression and anxiety disorders, with increasing rates of self-harm; girls were particularly vulnerable. I was told that the mental health problems were baked in when students arrived for ninth grade: Coming out of middle school, many students were already anxious and depressed. Many were also already addicted to their phones.

Ten months later, I was invited to give a talk at my old middle school. There, too, I met with the principal and her top administrators, and I heard the same thing: Mental health problems had recently gotten much worse. Even many of the students arriving for sixth grade, coming out of elementary school, were already anxious and depressed. And some were already addicted to their phones.[40]

We need to start prevention early, in elementary and middle schools, *before* our children begin wilting. Phone-free and play-full schools are easy to implement in those grades, and they cost very little money, especially when compared with the standard approach of hiring more therapists and buying new curricula.[41]

Let's test the two whales experimentally so that we learn whether these approaches work and which variations work best. And let's do it using entire schools for the interventions so that we can examine changes in school culture, rather than using individual children or individual classes within a single school.[42]

Here's how it might work: A school district superintendent, or a state-level education commissioner, or a governor—anyone who has influence with at least a few dozen elementary and middle schools—would recruit a pool of interested schools. Those schools would then be randomly assigned[43] into four experimental groups: (1) phone-free, (2) play-full (that is, Play Club plus extra recess), (3) phone-free plus play-full, and (4) the control condition, in which each school carries on with whatever it was doing before, but is asked not to change phone or recess policies.[45] In just two years, we'd find out whether these interventions work, whether one of them is stronger than the other, and whether there is an added benefit to combining them.

There are many variations of this basic experiment, adding or subtracting conditions or implementing policies in different ways.[45] The Let Grow Project could be included as part of the play-full school condition, because it draws on and amplifies the autonomy, risk-taking, and independence fostered by free play. Or a study could simply compare schools that do the Let Grow Project with those that don't.

Anxious and depressed students have been flowing from middle schools into high schools since the early 2010s, and high schools are struggling to respond—as are universities. But we can stem the flow. If we can keep smartphones entirely out of elementary and middle schools while making more room for free play and student autonomy, then the students who enter high school in a few years will be healthier and happier. If schools take these steps, in concert with parents taking related steps at home and governments changing laws to support those efforts, then I believe we can reverse the surge of suffering that hit adolescents in the early 2010s.

IN SUM

- U.S. middle and high schools have seen an increase in mental illness and psychological suffering among their students since the early 2010s. Many are implementing a variety of policies in response.

- There is a Polynesian expression: "Standing on a whale, fishing for minnows." Sometimes what you are looking for is right there, under-foot, and it is better than anything you could find by looking farther away. I suggested two potential whales that schools can implement right away, with little or no additional money: going phone-free, and becoming more play-full.

- Most schools say they ban phones, but that typically means only that students must not use their phones during class. This is an in-effective policy because it incentivizes students to hide their phone use during class and increase their phone use after class, which makes it harder for them to form friendships with the kids around them.

- A better policy is to go phone-free for the entire school day. When students arrive, they should put their phones into a dedicated phone locker or into a lockable phone pouch.

- The second whale is becoming a play-full school. The simple addi-tion of a Let Grow Play Club—an afternoon option in K–8 schools of playing on the school playground, with no phones, plenty of loose parts, and minimal adult supervision—may teach social skills and reduce anxiety better than any educational program, because free play is nature's way of accomplishing these goals.

- Schools can become more play-full by improving recess in three ways: Give more of it, on better playgrounds (such as those incorpo-rating loose parts and "junk," and/or more natural elements), with fewer rules.

- The Let Grow Project is another activity that seems to reduce anxi-ety. It is a homework assignment that asks children to "do some-thing they have never done before, *on their own*," after reaching agreement with their parents as to what that is. Doing projects increases children's sense of competence while also increasing parents' willingness to trust their children and grant them more autonomy.

- When all the families in a neighborhood or town give their children more free play and independence, it solves the collective action problem: Parents are no longer afraid to give their children more unsupervised free play and independence, which children need to overcome normal childhood anxieties and develop into healthy young adults.

- Schools can do more to reverse the growing disengagement of boys and their declining academic progress relative to girls. Offering more shop classes and more vocational and technical training and hiring more male teachers would each serve to re-engage boys. (As would offering better recess in the earlier years.)

- An ounce of prevention is worth a pound of cure. If K–8 schools become phone-free and play-full, and if they add in the Let Grow Project, they will be applying many pounds of prevention, which will reduce the flow of depressed and anxious students entering high schools.

Chapter 12

WHAT PARENTS CAN DO NOW

I n *The Gardener and the Carpenter*, the developmental psychologist Alison Gopnik notes that the word "parenting" was essentially never used
until the 1950s, and only became popular in the 1970s. For nearly all of
human history, people grew up in environments where they observed
many people caring for many children. There was plenty of local wisdom
and no need for parenting experts.

But in the 1970s, family life changed. Families grew smaller and more
mobile; people spent more time working and going to school; parenthood
was delayed, often into the 30s. New parents lost access to local wisdom
and began to rely more on experts. As they did so, they found it easy to
approach parenthood with the mindset that had led them to success in
school and work: If I can just find the right training, I can do the job well,
and I'll produce a superior product.

Gopnik says that parents began to think like carpenters who have a
clear idea in mind of what they are trying to achieve. They look carefully
at the materials they have to work with, and it is their job to assemble
those materials into a finished product that can be judged by everyone

against clear standards: Are the right angles perfect? Does the door work? Gopnik notes that "messiness and variability are a carpenter's enemies; precision and control are her allies. Measure twice, cut once."[1]

Gopnik says that a better way to think about child rearing is as a gardener. Your job is to "create a protected and nurturing space for plants to flourish." It takes some work, but you don't have to be a perfectionist. Weed the garden, water it, and then step back and the plants will do their thing, unpredictably and often with delightful surprises. Gopnik urges us to embrace the messiness and unpredictability of raising children:

> Our job as parents is not to make a particular kind of child. Instead, our job is to provide a protected space of love, safety, and stability in which children of many unpredictable kinds can flourish. Our job is not to shape our children's minds; it's to let those minds explore all the possibilities that the world allows. Our job is not to tell children how to play; it's to give them the toys. . . . We can't make children learn, but we can let them learn.

In this book I have argued that we have vastly and needlessly overprotected our children in the real world. In Gopnik's terms: Many of us have adopted an overcontrolling carpenter mentality, which prevents our children from flourishing. At the same time, we have underprotected our children in the virtual world by leaving them to their own devices and failing to do much weeding. We let the internet and social media take over the garden. We have left young people to grow up in digital social networks rather than in communities where they can put down roots. Then we are surprised that our children are lonely, starving for real human connections.

We need to become thoughtful gardeners in both realms. In the following pages, Lenore and I offer specific suggestions for how to do that, organized by the age of the child (although some of the suggestions apply to more than one age).[2]

FOR PARENTS OF YOUNG CHILDREN (AGES 0 TO 5)

In the first few years of life, children are developing basic perceptual and cognitive systems (such as vision, hearing, and language processing) and mastering basic skills (walking, talking, fine motor skills, agility skills like climbing and running). In the early years, as long as the child has a "good enough" environment with good nutrition, loving adults, and time to play, there is a limit to what parents can do to create better-than-normal outcomes.[3] What young children need is a lot of time to interact with you, with other loving adults, with other kids, and with the real world. Particularly in these years, and particularly in the United States, child care is an enormous and vexing puzzle. But these are the larger goals to keep in mind.

More (and Better) Experience in the Real World

As I discussed in chapter 3, attachment theory tells us that children need a secure base—a reliable and loving adult who will be there for them when needed. But the function of that base is to be a launching point for adventures off base, where the most valuable learning happens. Many of the best adventures are going to happen with other children, in free play. And when that play includes kids of mixed ages, the learning is deepened because children learn best by trying something that is just a little beyond their current abilities—in other words, something a slightly older kid is doing. Older kids can also benefit from interacting with younger kids, taking on the role of a teacher or older sibling. So, the best thing you can do for your young children is to give them plenty of playtime, with some age diversity, and a secure loving base from which they set off to play.

As for your own interactions with your child, they don't have to be "optimized." You don't have to make every second special or educational. It's a relationship, not a class. But what you *do* often matters far more than what you *say*, so watch your own phone habits. Be a good role model who is not giving continuous partial attention to both the phone and the child.

Also, trust young children's deep desire to help out. Even at age 2 or 3, children can put the forks on the place mats or help load the washing machine. Providing children with responsibility around the house makes them feel like an essential part of the family, and giving them more responsibility as they grow could offer some protection against later feelings of uselessness. In fact, a rising number of adolescents now agree with the statement "My life is not very useful."[4]

Less (and Better) Experience on Screens

Smartphones, tablets, computers, and televisions are not suitable for very young children. Compared with other objects and toys, these devices transmit intense and gripping sensory stimulation. At the same time, they encourage more passive behavior and information consumption, which can delay learning. This is why most authorities recommend against making screens a part of daily life in the first two years, and using them sparingly until the age of 6 or so.[5] The child's brain is "expecting" to wire up in a three-dimensional, five-senses world of people and things.

But one kind of screen time may be valuable in moderation: interacting with family members or friends via FaceTime, Zoom, or other video platforms. A key insight gained from research on screens and young children is that *active*, synchronous virtual interactions with other humans—what most of us call a video chat—can foster language learning and bonding, while *passive*, asynchronous viewing of a prerecorded video yields minimal benefits and in some cases even backfires and disrupts language learning, particularly for those under 2 years old.[6]

Expert advice on screens is clear and somewhat consistent across the Western world.[7] A representative set of recommendations that seem reasonable to me comes from the American Academy of Child & Adolescent Psychiatry:[8]

- Until 18 months of age, limit screen use to video chatting along with an adult (for example, with a parent who is out of town).

- Between 18 and 24 months, screen time should be limited to watching educational programming with a caregiver.
- For children 2–5, limit noneducational screen time to about one hour per weekday and 3 hours on the weekend days.
- For ages 6 and older, encourage healthy habits and limit activities that include screens.
- Turn off all screens during family meals and outings.[9]
- Learn about and use parental controls.
- Avoid using screens as pacifiers, babysitters, or to stop tantrums.
- Turn off screens and remove them from bedrooms 30–60 minutes before bedtime.

I understand how hard it is to raise young children without the assistance of a screen to keep them occupied and quiet while you make dinner or take a work call, or when you just need a break. My wife and I used the television show *Teletubbies* to mesmerize and calm our children from infancy through the toddler years. But if we had it to do over again, we'd do less of it.

FOR PARENTS OF CHILDREN AGES 6–13 (ELEMENTARY AND MIDDLE SCHOOL)

Once basic skills have been mastered in early childhood, children need to head out and take on more advanced challenges, including social challenges. They begin to care more about social norms and about how their peers see them. Compared with the preschool years, shame and embarrassment become more common and painful.[10] Children and adolescents in this age range are going through sensitive periods for cultural learning and risk assessment. It is during elementary school that children start applying the learning mechanisms I described in chapter 2 in full force: conformist bias (do what everyone else is doing) and prestige bias (copy the people whom everyone else looks up to). Because of children's ravenous appetite for social learning, it is important for parents to think about who those healthy learning models will be, and how to get them into a child's life.

More (and Better) Experience in the Real World

To make this period of social learning go well, it is best if children and young adolescents get a lot of experience by doing things together (such as in play), or with adults in their real-world communities, rather than sitting alone watching videos, playing mobile games, or reading posts from people online.

To reduce overprotection in the real world and encourage more productive off-base adventures, consider the following seven suggestions from Lenore.

1. **Practice letting your kids out of your sight without them having a way to reach you.** While you cook dinner for your friends, send your kids out with theirs to the grocery store to pick up more garlic (even if you don't need it). It is only by letting your kids out of your sight, untethered, that you will come to see that this is doable, and rather great. (It is probably what you were doing by the time you were eight.) This kind of practice will help you feel more prepared to give them more independence *and* to hold off on giving them a phone because you'll have seen for yourself that they can do fine without one. Just give the kids a note that they can show adults that says they have your permission to be out without you. You can print out such a card at LetGrow.org that says "I'm not lost or neglected!" and includes your phone number.[11]

2. **Encourage sleepovers, and don't micromanage them,** although if the friend brings a phone, hold on to it until the friend leaves, otherwise they'll have a phone-based sleepover.

3. **Encourage walking to school in a group.** This can begin as early as first grade if the walk is easy and there is an older child to be responsible. Safe Routes to School[12] is an organization that can help make sure there are stop signs, bike paths, crossing guards, and so on that allow kids to get to school. If school is too far to walk or ride a bicycle, consider "drive to five." You drop kids off, with other kids, at a spot that is five minutes away from school. Let them walk that last leg by themselves. (Schools can help organize this. It also reduces traffic congestion near the school.)

4. **After school is for free play.** Try not to fill up most afternoons with adult-supervised "enrichment" activities. Find ways that your children can just hang out with other children such as joining a Play Club (see chapter 11), or going to each other's homes after school. Friday is a particularly good day for free play because children can then make plans to meet up over the weekend. Think of it as "Free Play Friday."

5. **Go camping.** At campgrounds, kids are usually way more free range than at home, for a few reasons. First, they are away from their scheduled activities. Second, they're living in a small space with their parents—the outdoors beckons! Third, kids at campsites are *expected* to run around with the other kids. If you don't like camping, consider taking your next trip with another family that shares your ideas about independence, so the kids can play together.

6. **Find a sleepaway camp with no devices and no safetyism.** Many summer camps offer children and adolescents the chance to be out in nature and away from their devices and the internet for a month or two. Under those conditions, young people attend fully to each other, forming friendships and engaging in slightly risky

and exciting outdoor activities that may bond them together tightly. Avoid camps that are essentially summer school, with academic work and internet access, or camps that do not provide children with any communal responsibilities. Try to find a camp that embraces the values of independence and responsibility.[13] If possible, send your child there every summer, from third or fourth grade through eighth or ninth grade—or all the way through high school if they want to transition from camper to counselor.[14] Bonus points for any camp that promises to *not* post pictures every day on its website. Summer camp is a great opportunity for parents and children to get out of the habit of constant contact and, especially for parents, constant reassurance that their kids are okay.

7. **Form child-friendly neighborhoods and playborhoods.** Even if your neighborhood feels empty today, that doesn't mean it has to stay that way. If you can find one other family to join you, you can take simple steps that will activate common desires among neighbors and reanimate a block or neighborhood. Organize a block party with some activities for children. Then turn the neighborhood into a "playborhood,"[15] a term coined by Silicon Valley dad Mike Lanza, who made his yard into a gathering place for the local kids. If you'd like to do that too, he suggests you invite some neighbors over for lunch outside and put some things out for the kids to play with—a giant cardboard box, hula hoops—and they'll start playing. Announce that you'd be happy to have kids come by any Friday afternoon, or whatever time frame works for you. The key is regularity: Kids will come if they know other kids will be there.[16] *Another option:* Have parents take turns sitting outside one afternoon a week. Families can send their kids out knowing an adult is present for emergencies.

When I first read *Free-Range Kids*, I knew, as a psychologist, that Lenore's advice was sound. But every time my wife, Jayne, and I took her advice over the following years, we had to overcome our own palpable anxiety. The first time we let our son, Max, walk to school by himself in fourth grade (after several days of me "tailing" him, 20 yards behind), we held our breath as we watched his little blue dot pause at busy Seventh Avenue and then slide across it. (We gave him an iPhone when he started walking to school. Knowing what we know now, we'd have given him a phone watch or basic phone.) The first time our daughter brought lunch to me at my office, when she was 6, she had to cross one street, which was scary for me but thrilling for her. She looked both ways five times (I peered around the corner of my office building) and then came running into the lobby, jumping up and down with so much excitement that my lunch practically flew out of her delivery bag.

The cure for such parental anxiety is exposure. Experience the anxiety a few times, taking conscious note that your worst fears did not occur, and you learn that your child is more capable than you had thought. Each time, the anxiety gets weaker. After five days of our son walking to school, we stopped watching his blue dot. We got more comfortable with his ability to navigate the city, and soon its subway system. In fact, one of the pivotal moments in Max's development could never have happened if we had not followed Lenore's advice for years beforehand. When he was 12, Max had become very interested in tennis, and I had taken him to the U.S. Open, in Queens, a 40-minute subway ride from our apartment, with one transfer. The next year, when he was 13, he wanted to go to a particular night match by himself. Jayne and I were hesitant, but Max assured us that he could do it, and he really did know the subway system better than we did. So we conjured up an image of Lenore in our minds, and we said okay.

Max had a fantastic time at the match, which ran past 11:00 p.m. No problem, he flowed with the boisterous crowd to the nearby subway afterward. The problem arose at the transfer station; the train Max needed for the last mile home was not running that night. Max was nervous, but he improvised. He walked upstairs from the subway station and hailed a yellow cab—which I had taught him how to do but he had never done on his own—and made it home safely at 1:00 a.m. From that day on he was a different person, with more confidence, and from that day on we treated him differently and gave him still more independence. Jayne and I would not have said yes to Max's request had we not let him walk to school years earlier, and grown to trust him without tracking his blue dot at every moment.

Less (and Better) Experience on Screens

Children in elementary and middle school are doing a lot of learning, and screen-based activities can play a valuable role. However, for many children, time spent on screens expands like a gas to fill every available moment, and the content of that gas is almost entirely entertainment, not educational. So it is not enough to just delay the first smartphone until high school; parents need to keep a lid on total screen-based

"It's been so nice getting to interact with you for these past six years. Here's your first device."

Figure 12.1. H. Lin, in *The New Yorker*.[17]

activities because of the high opportunity cost that they impose and the habits they create. Parents should also be mindful of the behaviors they model.[18]

The average 8-to-12-year-old spends between four and six hours a day on recreational screen activities, across multiple screens.[19] This is why most medical authorities and national health organizations recommend that parents place a limit on total recreational screen time for children in this age range. The government of Quebec offers representative guidance in a concise form with the right level of flexibility:

> **For 6-to-12-year-olds:** As a general rule, no more than two hours per day is recommended for screen-based recreational activities. However, this depends on the content (social media, video games, chats, TV, and so forth), the context (time of day, multitasking, and so on), and the young person's individual traits (age, physical and mental health, analytical skills, critical thinking, and so on). Parental supervision must therefore be based on these criteria. For younger children especially, the content should be educational, and the devices should be used in common areas, where adults can control the content, rather than in children's bedrooms.

Drawing from the various lists of recommendations, and from the research presented earlier in this book, I offer the following additional suggestions:

1. **Learn how to use parental controls and content filters** on all the digital devices in your home. You want your children to become self-governing and self-controlled, with no parental controls or monitoring by the time they reach age 18, but that does not mean you should immediately give them full independence in the online world before their frontal cortex is up to the task. Tech companies employ tools that will hook children, so use parental controls in this age range to fight back. And if it makes sense for your family,

set a total amount of time for recreational screen use. Time limits can be complicated to work out, and they can backfire if set too high (the child will then try to "use up" all available time[20]). But if you don't set a total limit, the platforms will grab more and more time, including sleep time. Some parents use monitoring programs that allow them to read their children's texts and other communications. There may be cases in which this is necessary, but in general I think it is preferable to avoid monitoring private conversations and to focus instead on blocking access to age-inappropriate sites and apps, and specifying times when devices can and cannot be used. It is possible to overprotect in the virtual world, especially when it shades over into surveillance, sometimes without the child's knowledge. Visit CommonSenseMedia.com for guidance on using parental controls.[21]

2. **Focus more on maximizing in-person activity and sleep than on total screen hours.** The main harm done by most screen activities is the opportunity cost, which directly drives two of the four foundational harms that I described in chapter 5: social deprivation and sleep deprivation. If your children are spending a lot of time in person with friends, such as on sports teams or in unstructured play or hangouts, if they are getting plenty of sleep, and if they show no signs of addiction or problematic use on any devices, then you may be able to loosen up on the screen-time limit. Likewise, playing video games with a friend, in person and in moderation, is better than playing alone in one's room. Leonard Sax, author of *Boys Adrift*, recommends no more than 40 minutes a night on school nights, and no more than an hour a day on weekends.[22] However, many families use the rule of allowing longer periods, but only on weekends. As with social media use, limits are hard to impose if you are the only family imposing them, so try to coordinate with the parents of your child's friends. When many families impose similar limits, they break out of the collective action trap and everyone is better off.

3. **Provide clear structure to the day and the week.** As we saw in chapter 8, structuring time and space is a precondition for rituals and other communal activities, which strengthen the feeling of belonging in a community—even one as small as a two-person family. Shared meals should be phone-free so that family members attend to each other. Having a regular family movie night would be good. Be wary of allowing devices in bedrooms at these younger age, but if you do, then all devices should be removed from bedrooms by a fixed time, which should be at least 30 minutes before the scheduled bedtime.[23] Consider taking a "digital Sabbath" every week: a full day where no screen devices are used. Consider taking a screen-free week every year, perhaps on a vacation in a beautiful natural setting.

4. **Look for signs of addiction or problematic use.** Screen-based activities are fun, and video games in particular are widely enjoyed by nearly all children in this age range. As I showed in chapter 7, video games in moderation do not seem to be harmful for most children, and yet there is a large subgroup of children and adolescents (in the ballpark of 7%) who end up either truly addicted or else showing signs of what is called problematic use, which means that the activity is interfering with other areas of functioning. Pornography, social media, and video games are the three categories of activity most likely to lead to problematic use among adolescents, and years of problematic use may cause lasting changes, as Chris discussed in chapter 7 when he said he feels like a "hollow operating system." The American Psychological Association offers these guidelines for recognizing "problematic social media use," but they apply fairly well to any screen-based activity.

Your child's social media use might be causing problems if:

- it interferes with their daily routines and commitments, such as school, work, friendships, and extracurricular activities

- they experience strong cravings to check social media
- they lie or use deceptive behavior to spend time online
- they often choose social media over in-person social interactions
- it prevents them from getting at least eight hours of quality sleep each night
- it prevents them from engaging in regular physical activity
- they keep using social media even when they express a desire to stop

If your children show one or more of these signs, you should talk to them. If they can't self-correct immediately, or if they are showing multiple signs, then take steps to remove access for a period of time to allow for a digital detox and dopamine reset. Consult sites that specialize in advice for video game and social media addiction.[24]

5. **Delay the opening of social media accounts until 16.** Let your children get well into puberty, past the most vulnerable early years, before letting them plug into powerful socializing agents like TikTok or Instagram. This doesn't mean they can never see any content from these sites.; as long as they can get to a web browser, they'll get to the platforms. But there's a difference between *viewing* TikTok videos on a browser and *opening an account* on TikTok, which you reach via the app on your smartphone during every spare moment. Opening an account is a major step in which adolescents provide personal data to the platform, put themselves into a stream of personalized content chosen by an algorithm to maximize engagement, and begin to post content themselves. Delay that fateful step until well into high school.

6. **Talk with your preteen about the risks, and listen to their thoughts.** Even without a social media account, all children will encounter age-inappropriate content online. Exposure to pornography is virtually certain. Talk with your preteens about the risks inherent in

posting public content or sharing personal information online, including sexting and cyberbullying. Ask them what problems they see in their peers' online habits, and ask them how they think they can avoid such problems themselves.[25]

You have to let go online eventually. But if you can keep the quantity of online time lower and the quality higher in this long period of childhood and early adolescence (ages 6–13), you'll make room for more real-world engagement, and you'll buy time for your child's brain to develop better self-control and less fragmented attention.

FOR PARENTS OF TEENS AGES 13–18 (HIGH SCHOOL)

Consistent with the notion of a path to adulthood, the transition to high school should be a major milestone at which adolescents get an increase in freedoms and responsibilities in the real world and in the virtual world.

More (and Better) Experience in the Real World

Adolescents have nearly all begun puberty by the time they start high school, and this is the period when rates of depression and anxiety start to rise more steeply. In earlier chapters, I made the case that helping young people feel useful and connected to real-world communities is pivotal to their social and emotional development, so it is important that adolescents take on some adult-level challenges and responsibilities. Finding non-parental role models also becomes more valuable during this period.

1. **Increase their mobility.** Let your teens master the transportation modes that make sense for where you live: bicycles, buses, subways, trains, whatever. As they grow, so should the boundaries of their world. Encourage them to get their driver's licenses as soon as they

are eligible, and give them driving lessons and encouragement to use the car, if you have one. Encourage more and better off-base excursions with friends. Let your teen hang out at a "third place" (not home or school) like the Y, the mall, the park, a pizzeria—basically, a place where they can be with their friends, away from adult supervision. Otherwise, the only place they can socialize freely is online.

2. **Rely more on your teen at home.** Teens can cook, clean, run errands on a bicycle or public transit, and, once they turn 16, run errands using a car. Relying on your teen is not just a tool to instill work ethic. It's also a way to ward off the growing feeling among Gen Z teens that their lives are useless. One 13-year-old told Lenore that when she started doing more things on her own, including runs to the drugstore for her mom, and getting herself places without being driven, she started to realize just how much time her mom spent doing boring, thankless things like carpooling and sitting through freezing soccer games. Once she started empathizing with her mom—and helping out more—the two stopped fighting as much because, in a way, now they were on the same team.

3. **Encourage your teen to find a part-time job.** Having a boss who is not a parent is a great experience, even when it's not a pleasant one. Even one-off gigs are good. Shoveling a neighbor's driveway requires talking to an adult, negotiating a fee, and completing the task. Earning your own money—and having control over how it is spent—is an empowering feeling for a young person.

4. **Find ways for them to nurture and lead.** Any job that requires guiding or caring for younger children is ideal, such as a babysitter, camp counselor, or assistant coach. Even as they need mentors themselves, they can serve as a mentor to younger kids. Helping younger kids seems to turn on an empathy switch and a leadership gene. Lenore saw this happen the first time her younger son went on

a Boy Scout overnight camping trip, at the age of 11. He was beyond excited. He was also beyond unprepared: He forgot his sleeping bag. Oh, did he cry when he realized that; he thought he'd be sent home. Then an older Scout—a high school student—said, "Don't worry! I always bring an extra sleeping bag for just this kind of situation!" Lenore's son was grateful, and so was she when she heard the story. Lenore was even more grateful years later when she learned that in fact the older Scout had *not* brought an extra sleeping bag. He slept on the cold, hard ground. That's how you become a leader.

5. **Consider a high school exchange program**. These have a long history. A visitor to England in 1500 wrote, "Everyone, however rich he may be, sends away his children into the houses of others, whilst he, in return, receives those of strangers into his own."[26] Yes, even in medieval Britain, people realized this experience would broaden a kid's world. It can also be easier for a kid to listen to someone other than Mom or Dad. One modern-day program to consider is the American Exchange Project.[27] It sends high school seniors from all over the United States to spend a week with a family in another state, in the hopes of weaving a polarized country back together. And it's free! Meanwhile, the American Field Service has been sending high school students all over the globe for decades.[28] Teens live with a family and attend the local school. Alternatively, you can host a student from abroad.[29] CISV International, pioneered by the child psychologist Dr. Doris Allen, fosters intercultural friendship through exchanges and other youth programming beginning at age 11. There are CISV chapters in more than 60 countries around the world.[30]

6. **Bigger thrills in nature.** Let your teens go on bigger, longer adventures, with their friends or with a group: backpacking, rock climbing, canoeing, hiking, swimming—trips that get them out into nature and inspire real-world thrills, wonder, and competence. Consider programs that run a month or longer with organizations such as

Outward Bound and the National Outdoor Leadership School, which are designed to foster self-reliance, social responsibility, self-confidence, and camaraderie (and do not require prior outdoor experience). There are also a number of free or subsidized programs,[31] as discussed in chapter 10.[32] As Kurt Hahn, the founder of Outward Bound, explained,

> There is more to us than we know. If we can be made to see it, perhaps for the rest of our lives we will be unwilling to settle for less. There exists within everyone a grand passion, an outlandish thirst for adventure, a desire to live boldly and vividly through the journey of life.

7. **Take a gap year after high school.** Many young people go directly to college without any sense of what else is out there. How are they supposed to know what they want to do with their lives—or even whether college is their best option? Let young adults discover more about their interests and about the world. They can get a job and save up money. Travel. Volunteer. They are not damaging their college prospects. They are improving their chances of finding a path they want to pursue, and they are improving their competence at following any path. A gap year is intended not to postpone a young person's transition to adulthood but rather to accelerate it. It's a year to build skills, responsibility, and independence. You can find a list of organizations that can help your teen plan a gap year at gapyearassociation.org. Scholarships and subsidies are often available.[33]

The idea behind all of these suggestions is to let teens grow more confident and competent by engaging with the real world. Encourage activities that stretch them beyond their comfort zone. Yours too! Risking a serious injury for no good reason is dumb. But *some* risk is part of any hero's journey, and there's plenty of risk in not taking the journey too.

Less (and Better) Experience on Screens

The teen years should be a time of loosening restrictions as teens mature and gain greater ability to inhibit impulses and exercise self-control. The frontal cortex is not fully developed until the mid-20s, but a 16-year-old can and should be given more autonomy and self-determination than a 12-year-old.

Whenever you transition your teens from basic phones to smartphones, talk to them and monitor how the transition is going. You should continue to set parameters within which they have autonomy, such as maintaining family rules about when phones and other devices can and cannot be used. High school students are even more likely to be sleep deprived than middle school students, so help your teen develop a good evening routine, one in which the phone is removed from the bedroom by a set time each night. Most of my students say that the last thing they do at night before closing their eyes is to check their texts and social media accounts. It's also the first thing they do in the morning before getting out of bed. Don't let your children develop this habit.

Whenever you let your children open social media accounts, look out for signs of problematic or addictive use. Ask them how their online lives are helping them to achieve their goals, or hindering them. Educate them about how social media works and how it hooks and harms many users by going through the "Youth Toolkit" and other resources at the Center for Humane Technology.[34]

I WANT TO MAKE ONE FINAL POINT, ABOUT HOW SMARTPHONES CHANGED parent-child relationships. Around 2012, when adolescents started getting smartphones—and when those anxiety graphs shot upward—something else happened: Their parents got smartphones too. Those smartphones gave parents a new superpower that they did not have in the era of flip phones: the ability to track their children's movements at every moment. Lenore pointed out to me that this could be part of the reason for increased anxiety and decreased confidence. Parents began to surveil their children everywhere, such as when they're on their way to school, or when they're hanging out with their friends after school. If anything looked unusual, the parent could call or text right away, or could grill the child about her activities when she returned home. Whether we think of the phone as "the world's longest umbilical cord" or as an "invisible fence," childhood autonomy plummeted when kids started carrying them. Even if a parent rarely looks at the tracker, and even if a kid never summons Mom to come get him because his bike chain broke, the fact that this is always possible makes it more difficult for children and adolescents to feel that they are on their own, trusted and competent. And it makes it more difficult for parents to let go.

Lenore and I have debated the merits of tracking for years. Jayne and I began tracking our children as soon as we gave them phones, and we know it made it easier for us to let them out earlier to begin their free-range childhoods in New York City. But as I have heard Lenore describe the growing surveillance of children and the computer-assisted monitoring of their academic performance, sometimes with instant notification of grades and daily updates on classroom behavior, I have begun to feel creeped out. And even though tracking helped Jayne and me gain confidence in our kids when they were young, and it helps us now to manage family logistics, such as when everyone will get home for dinner, will we ever turn it off? Should we? I don't know the answer.

IN SUM

- Being a parent is always a challenge, and it has become far more challenging in our era of rapid social and technological change. However, there is a lot that parents can do to become better "gardeners"— those who create a space in which their children can learn and grow—in contrast to "carpenters" who try to mold and shape their children directly.

- If you do one thing to be a better gardener in the real world, it should be to give your children far more unsupervised free play, of the sort you probably enjoyed at that age. That means giving them a longer and better play-based childhood, with ever-growing independence and responsibility.

- If you do one thing to be a better gardener in the virtual world, it should be to delay your children's full entry into the phone-based childhood by delaying when you give them their first smartphone (or any "smart" device). Give only basic phones before the start of high school, and try to coordinate with other parents so that your children do not feel that they are the only ones without smartphones in middle school.

- There are many other ways to increase your children's engagement with the real world and embeddedness in communities, including sending them to a technology-free sleepaway camp, going camping, and helping them find additional settings in which they can hang out with other children who are not carrying smartphones.

- As your children get older, increase their mobility and encourage them to find part-time jobs and ways to learn from other adults. Consider an exchange program, a summer wilderness program, and a gap year.

- A free-range childhood is more likely to produce confident, competent young adults, with lower levels of anxiety, than is a childhood ruled by safetyism, fear, and constant adult supervision. The biggest obstacle is the parents' own anxiety about letting a child out of sight, unchaperoned by an adult. This takes practice, but the ultimate pleasure of being able to trust your child outweighs the temporary anxieties of letting go.

- Most authorities recommend little or no screen time in the first 18–24 months (other than video calls with family members) and limited screen time through the age of 5 or 6.

- For children in elementary and middle school, use parental controls, provide clear limits, and designate some times and places as no-device zones. Look out for signs of problematic or addictive use.

- Your actions as a parent can contribute to solving the collective action problem. If you delay giving your child a smartphone, it makes it easier for other parents to do so. If you give your child more independence, it makes it easier for other parents to do so too. If you do it together, with other families, it will make it easier still, and more fun.

Conclusion

BRING CHILDHOOD
BACK TO EARTH

didn't set out to write this book. In late 2021, I began writing a book on how social media was damaging American democracy. My plan was to begin with a chapter on the impact of social media on Gen Z, showing how it disrupted their social lives and caused a surge of mental illness. The rest of the book would analyze how social media disrupted society more broadly. I'd show how it fragmented public discourse, Congress, journalism, universities, and other foundational democratic institutions.

But when I finished writing that first chapter—which became chapter 1 of this book—I realized that the adolescent mental health story was so much bigger than I had thought. It wasn't just an American story, it was a story playing out across many Western nations. It wasn't just about girls, it was about boys too. And it wasn't just about social media. It was about the radical transformation of childhood into something inhuman: a phone-based existence.

As Zach and I compiled studies into a dozen review documents, including one that collected evidence for all the *other* theories about what could be causing the mental health crisis,[1] we grew more confident that the key culprit was the rapid transformation of childhood that happened between 2010 and 2015. Many other factors contributed to the mental health challenges faced by Gen Z in the United States, but no other theory could explain why similar difficulties arose for Gen Z in so many other nations at the same time. Until someone finds a chemical that was released in the early 2010s into the drinking water or food supply of North America, Europe, Australia, and New Zealand, a chemical that affects adolescent girls most, and that has little effect on the mental health of people over 30, the Great Rewiring is the leading theory.

I decided to split the book I was writing into two books, and to write this one first because the adolescent mental health crisis is so urgent and because there is so much that we can do, today, to turn it around. As a social scientist, as a teacher, and as a father of two teens, I don't want to wait. I want us to get moving. If the phone-based childhood is a major contributor to the international epidemic of mental illness, then there are a few clear and powerful actions that parents, teachers, and members of Gen Z can take to roll back the phone-based childhood.

In part 4, I offered dozens of suggestions, but the four foundational reforms are:

1. No smartphones before high school
2. No social media before 16
3. Phone-free schools
4. Far more unsupervised play and childhood independence

These reforms are foundational because they solve multiple collective action problems. Each parent who takes action makes it easier for other parents in the community to do the same. Each school that goes truly phone-free liberates all of its students to be more present with each other. If a community enacts all four, they are likely to see substantial improvements in child and adolescent mental health within two years.[2]

As for how to bring about these reforms, I end this book with two suggestions: *Speak up* and *link up*.

SPEAK UP

In a classic 1968 social psychology experiment, Bibb Latane and John Darley brought Columbia University students into the lab to take part in what the students thought was a discussion on problems of urban life.[3] The real experiment took place in the waiting room as the students filled out some preliminary questionnaires. After a few minutes, a strange smoke began to pour into the room through a vent. Would students get up and go tell someone, or would they just sit there passively and keep filling out questionnaires?

In the control condition, students were alone in the waiting room. In that condition, 75% took action, with half of the subjects leaving the room to find the experimenter within two minutes of noticing the smoke's appearance. (The experimenters monitored and filmed everything through a one-way mirror.)

In another condition, three students were brought into the waiting room one at a time and seated at separate desks. The experimenters wanted to know: Would having multiple people witnessing the smoke increase or decrease the likelihood that *anyone* would take action? The answer: It decreased it. Only three of the 24 students who were in that condition got up to report the smoke, and only one did so within the first four minutes, even though the smoke began to obstruct everyone's vision by then.

I should point out that the smoke was not smoke from a fire; it was titanium dioxide, used to create smoke screens.[4] This is crucial: Nobody in that room could understand what was going on. When there is ambiguity, people look to each other to see what everyone else is doing. Those cues help define the situation. Is it an emergency? If everyone else is just sitting there, then each individual comes to the conclusion that no, it's not an emergency.

The diffusion of digital technology into children's lives has been like smoke pouring into our homes. We all see that something strange is happening, but we don't understand it. We fear that the smoke is having bad effects on our children, but when we look around, nobody is doing much about it.

The most important lesson here is to speak up. If you think the phone-based childhood is bad for children and you want to see a return to play-based childhood, say so. Most people share your suspicion, but they are not sure what to do about it. Speak to your friends, your neighbors, your coworkers, your social media followers, and your political representatives.

If you speak up and support the four foundational reforms, you will inspire many others to join you. If you are a member of Gen Z, society urgently needs your voice. Your words will be the most powerful of all.

LINK UP

If you're a parent, connect with other parents who value play-based childhood and more childhood independence. There are many excellent organizations that bring parents together around this cause, including Let Grow, Outsideplay, and Fairplay.[5] There are also many excellent organizations that bring parents together and offer ideas and resources for delaying the phone-based childhood or making it less damaging. See Fairplay, the Center for Humane Technology, Common Sense Media, Screen Strong, and others that I'll list in the online supplement.[6] Talk with the parents of your children's friends. Most likely, they share your concerns, and if you act together to delay smartphones and social media, then it will be easier for you and your children to reject the phone-based childhood and choose real-world community instead.

If you have children in school, join with other parents to speak directly to the principal or head of the school. Urge him or her to implement the big ideas in chapter 12: going phone-free, encouraging more independence and responsibility, and adding a lot more free play. I can assure you that most principals, administrators, and teachers hate the phones, but they need to hear a lot of support from parents before they can make such a change.

If you're a teacher and you're fed up with the social chaos and learning disruption caused by smartphones and social media, link up. Talk to your fellow teachers and urge your school's leadership to reconsider not just the policy on phones, but on all devices that let students text each other or check social media while they are sitting in your class. You shouldn't have to compete for your students' attention with the entire internet. See if your school can coordinate a message to parents asking them to support change. If teachers speak with a unified voice and ask parents for help educating their children, the odds of success are high.

If you're a member of Gen Z, consider joining one of the organizations founded by members of your generation to bring about change. See, for example, the collaboration at Design It For Us.[7] They are a youth-led coalition that advocates for policy reforms to protect kids, teens, and young adults online. As cochair Emma Lembke put it in her testimony to a U.S. Senate committee, "While our stories may differ, we share the frustration of being portrayed as passive victims of Big Tech. We are ready to be active agents of change, building new and safer online spaces for the next generation."[8]

I OPENED THIS BOOK WITH THE FANCIFUL STORY OF A TECH ENTREPRE-neur who transports children away from Earth to grow up on Mars, without their parents' consent. It's unfathomable that we would ever let such a thing happen. But in some ways, we have. Our children may not be on Mars, but they're not fully present here with us, either.

Humanity evolved on Earth. Childhood evolved for physical playfulness and exploration. Children thrive when they are rooted in real-world communities, not in disembodied virtual networks. Growing up in the virtual world promotes anxiety, anomie, and loneliness. The Great Rewiring of Childhood, from play-based to phone-based, has been a catastrophic failure.

It's time to end the experiment. Let's bring our children home.

TO LEARN MORE

If you want to learn more about the topics covered in this book, I offer three main resources:

1. AnxiousGeneration.com

This is the main hub for resources related to *The Anxious Generation*. It has separate pages where I collect research and advice for parents, for schools, for Gen Z, and for readers interested in the spiritual practices I described in chapter 8. It will also have links to the next two resources.

2. The Online Supplement

Zach Rausch and I maintain a separate Google Doc for each chapter in this book. We give many additional graphs that couldn't fit in the book. We keep these pages updated with new research findings, and I report any errors I have made or any places where I have changed my mind. Zach includes links to the datasets he used to create most of the graphs in the book. You can find all of the online supplement files at anxiousgeneration.com/supplement.

3. The *After Babel* Substack

There is so much more I wanted to say in this book. There are so many other chapters I thought about writing. I will write short versions of such chapters on my Substack, *After Babel*. Please sign up at www.afterbabel.com. It's free. I'll have posts on these topics and many more:

Advice from Gen Z for Gen Z

Growing Up Under Constant Surveillance

What Universities Can Do Now

What Employers Can Do Now

How Social Media Affects Boys

How Pornography Affects Girls

How the Great Rewiring Changed Romantic Life

Why Were Religious Conservatives Less Affected by the Great Rewiring?

Limbic Capitalism: How market forces have incentivized addiction for hundreds of years

New technologies are going to be disrupting our lives at a faster rate every year. Please join me at *After Babel* to study what is happening, what it's doing to us, and how to raise flourishing children amid the confusion.

ACKNOWLEDGMENTS

This book was a team effort, so let me start by offering three teammates special awards.

The first goes to Zach Rausch, a young man I hired in 2020 as my general research assistant. Zach shared my passion for applying social psychology to complex social problems. He took the lead on two questions for which I needed answers: What is happening internationally? And what is happening to boys? By the time I started writing this book in the fall of 2022, Zach had become my thought partner and editor. For fourteen months we worked together intensively. He even devoted many late nights and weekends to write what we thought, at the outset, would be a short book. In that time Zach evolved from a second-year psychology grad student into a first-rate researcher and intellectual. I could not have written this book without him.

The second special award goes to Lenore Skenazy. Ever since I read her book *Free Range Kids,* Lenore has been my parenting muse, and she became a close friend as well. I reached out to Lenore for guidance about what I should say to parents in this book. She filled up a Google Doc with so many great ideas that I invited her to join me in writing chapter 12. And then chapter 11 on schools. And then also chapter 10 on what governments can do. If this book persuades parents, schools, and legislators to give kids more independence, it will be thanks to Lenore's many years of work on this issue as the president of Let Grow, and her enormous contributions to writing part 4 of this book.

The third special award goes to my editor at Penguin Press, Virginia Smith. Ginny has guided and improved my writing since 2016, when Greg Lukianoff and I began working with her on *The Coddling of the American Mind*. Ginny did deep editing on every chapter of *The Anxious Generation*, and, together with associate editor Caroline Sydney, they made the book come together, despite my difficulty with timeliness.

I am grateful to many other members of the team who played crucial roles in bringing this book into existence. Eli George is a Gen Z writer and intellectual who worked closely with me on the entire project, contributing qualitative research, creative ideas, and superb editing. Ravi Iyer, my friend and longtime collaborator at YourMorals.org, came through with advice and several key paragraphs in chapter 10, on what tech companies and governments can do. Chris Saitta handled all the notes and helped us to understand what boys are going through. Cedric Warny supported Zach in developing the databases needed for this book. Dave Cicirelli, my cool artist friend who did the illustrations for *All Minus One*, worked his magic again to create the cover of this book.

I sent the manuscript out to dozens of friends and colleagues in the summer of 2023 with a request to find errors and rough spots. Many of them came through and made the book better in a thousand ways. I thank: Trevor Agatsuma, Larry Amsel, Mary Aviles, John Austin, Michael Bailey, Barbara Becker, Arturo Bejar, Uri Bilmes, Samantha Boardman, Dave Bolotsky, Drew Bolotsky, Maria Bridge, Ted Brugman, Mariana Brussoni, Maline Bungum, Rowan Byrne, Camille Carlton, Haley Chelemedos, Carissa Chen, Jim Coan, Grace Coll, Jackson Davenport, Samantha Davenport, Michael Dinsmore, Ashlee Dykeman, Lucy Farey, Ariella Feldman, Chris Ferguson, Brian Gallagher, Peter Gray, Ben Haidt, Francesca Haidt, Max Haidt, Jennifer Hamilton, Melanie Hempe, Alexandra

Hudson, Freya India, Andrea Keith, Nicole Kitten, Sena Koleva, Bill Kuhn, Elle Laub, John Lee, Anna Lembke, Meike Leonard, Lisa Littman, Julia Lombard, Sergio A. Lopez, Mckenzie Love, Greg Lukianoff, Joy McGrath, Caroline Mehl, Carrie Mendoza, Jamie Neikrie, Evan Oppenheimer, Pamela Paresky, Yejin Park, Robbie Pennoyer, Maria Petrova, Kyle Powell, Matt Pulford, Fernando Rausch, Richard Reeves, Jayne Riew, Jeff Robinson, Tobias Rose-Stockwell, Arthur Rosen, Nima Rouhanifard, Sally Satel, Leonard Sax, Rikki Schlott, David Sherrin, Yvette Shin, Daniel Shuchman, Mark Shulman, Bennett Sippell, Ben Spaloss, David Stein, Max Stossel, Jonathan Stray, Alison Taylor, Jules Terpak, Jean Twenge, Cedric Warny, and Keith Winsten.

A few people on that long list rose to the level of super-editor, with detailed comments on every page: Larry Amsel, Grace Coll, Michael Dinsmore, Brian Gallagher, Nicole Kitten, McKenzie Love, Maria Petrova, Jayne Riew, Mark Shulman, and Ben Spaloss.

I am so fortunate to be a professor at New York University's Stern School of Business. Dean Raghu Sundaram and my department chair, Batia Wiesenfeld, have given me unwavering support in challenging times. Stern's Business and Society Program is an exciting place to study the ways that business is affecting and sometimes upending society.

My greatest thanks goes to my wife, Jayne Riew, with whom I first dreamed of children, and with whom I now share the joys of watching two of them make ever more ambitious off-base excursions.

NOTES

Introduction: Growing Up on Mars

1. Hamm et al. (1998); Milder et al. (2017).
2. Grigoriev & Egorov (1992); Strauss, M. (2016, November 30). We may finally know why astronauts get deformed eyeballs. *National Geographic.* www .nationalgeographic.com/science/article/nasa-astronauts-eyeballs-flattened -blurry-vision-space-science.
3. See, for example, Meta's response to Frances Haugen's revelations in the Face-book Files: Zuckerberg, M. (2021, October 5). Facebook. www.facebook.com /zuck/posts/10113961365418581. See also my rebuttal of Mark Zuckerberg's claim that the research shows that using Instagram is "generally positive for their mental health": Fridman, L. (2022, June 4). Jonathan Haidt: The case against social media. *Lex Fridman Podcast #291* (video). YouTube. www.youtube .com/watch?v=f0un-l1L8Zw&ab_channel=LexFridman.
4. Once boys grew old enough, other companies began to sink their hooks into them, including sports-betting platforms and dating apps.
5. See here to learn about COPPA: Jargon, J. (2019, June 18). How 13 became the internet's age of adulthood. *Wall Street Journal.* www.wsj.com/articles/how-13 -became-the-internets-age-of-adulthood-11560850201. In 2023, there was sud-denly a great deal of bipartisan interest in protecting children from social media, with notable efforts in California and Utah, and a variety of bills intro-duced in the U.S. Congress, which I'll discuss in chapter 10.
6. Thorn & Benenson Strategy Group (2021); Canales (2021, May 13). 40% of kids under 13 already use Instagram and some are experiencing abuse and sexual

solicitation, a report finds, as the tech giant considers building an Instagram app for kids. *Business Insider*. www.businessinsider.com/kids-under-13-use-facebook -instagram-2021-5.

7. In chapter 10, I will discuss the U.K.'s Age Appropriate Design Code, a version of which has been enacted in California too. Several U.S. states also passed age verification requirements and other regulations in 2023.

8. Drum, K. (2016). Lead: America's real criminal element. *Mother Jones*. www .motherjones.com/environment/2016/02/lead-exposure-gasoline-crime-increase -children-health/; Kovarik, B. (2021, December 8). A century of tragedy: How the car and gas industry knew about the health risks of leaded fuel but sold it for 100 years anyway. *Conversation*. theconversation.com/a-century-of-tragedy-how -the-car-and-gas-industry-knew-about-the-health-risks-of-leaded-fuel-but-sold -it-for-100-years-anyway-173395. See both articles for reviews of the history of leaded gasoline and its effects on brain development and later criminality. Paint and water pipes were additional sources of lead poisoning.

9. Pew Research identifies 1997 as the first year of Gen Z, but I believe 1997 is a bit too late; the new behaviors were clear in college students arriving on campus in 2014 See Parker & Igielnik (2020). Jean Twenge chose 1995 as the first year of "iGen." I split the difference between them and choose 1996 as the first year of Gen Z. Of course, generations are not separated by a bright line; nonetheless, they differ, as Twenge shows in her 2023 book, *Generations*.

10. Of course, AI is looking like it will change *everything*, so we are likely to see a new generation beginning in the 2020s. But since AI is likely to pull children even farther away from the real world, my prediction is that it will lead to even higher levels of anxiety if we don't act now to reverse the Great Rewiring of Childhood.

11. She lays this out in her book *Generations* (Twenge, 2023a). See also her earlier book *iGen* (Twenge, 2017).

12. See for this story Haidt, J., & Rose-Stockwell, T. (2019). The dark psychology of social networks. *Atlantic*. www.theatlantic.com/magazine/archive/2019/12/social -media-democracy/600763/. I note that Tumblr had introduced a "reblog" feature in 2007, but its effects were small compared to Twitter's "retweet" in 2009.

13. Steinberg (2023, Introduction).

14. Examples include the rise of trigger warnings, safe spaces, and bias response teams, all of which were discussed in the *Atlantic* essay.

15. Twenge, Martin, & Campbell (2018).

16. See my summary of the research: Haidt (2023, February).

17. Durocher, A. (2021, September 2). The general history of car seats: Then and now. *Safe Ride 4 Kids*. saferide4kids.com/blog/the-general-history-of-car-seats/.

18. Food and Drug Administration (2010).

19. Epictetus (1st–2nd century/1890, chapter 33). *The Enchiridion*.

20. Marcus Aurelius (161–180 CE/2002, book 3, chapter 4).

21. There has been a general rise in suicides among adults (ages 50 and above) in the United States, Canada, the U.K., and Australia since 2010, but these changes are generally smaller than the changes among younger populations (in relative terms). Importantly, the rises we see in the 2010s among adults were often preceded by decades of declining rates in the 1980s and 1990s. See Rausch & Haidt (2023, October).

22. See my essay with Eric Schmidt on how AI will supercharge four existing problems with social media: Haidt, J., & Schmidt, E. (2023, May 5). AI is about to make social media (much) more toxic. *Atlantic*. www.theatlantic.com/technology /archive/2023/05/generative-ai-social-media-integration-dangers-disinformation -addiction/673940/.

Chapter 1: The Surge of Suffering

1. Names and minor details have been changed to protect privacy.
2. The exception to this statement is suicide rates among American teens. Those rates generally declined in the early 2000s, hitting lows in 2007. The rates generally begin to rise in 2008, but they don't rise above where they had been in the early 2000s until after 2010. I will discuss suicide rates in a later section. If we look further back in time, we see that rates of depression, anxiety, and other disorders have been rising among American adolescents since the 1950s, with fluctuations. But there is nothing like the "hockey stick" increases in the early 2010s, which you'll see throughout this chapter and this book. See Twenge et al. (2010).
3. Data through 2021: Substance Abuse and Mental Health Services Administration. (2023).
4. Research note on demographic variation: Since 2010, an escalating trend in adolescent mental illness is evident across all groups in the United States, whether we look by sex, race, sexual orientation, or social class. Overall, Black teenagers have long had lower rates of anxiety, depression, self-harm, and suicide than their white counterparts, but both groups have seen a sharp increase since 2010, with larger absolute rises among white teens and larger relative (percentage) rises among Black teens (because they begin from a lower base rate). Data concerning social class is scarce, but depression follows similar trends across classes, with steep increases starting from 2010. Compared with heterosexual teens, LGBTQ teens report significantly higher rates in all the above measures. However, the evidence is inconclusive regarding the growth of self-harm and suicide rates among LGBTQ teens since 2010. For sources on these statistics and additional content, see the online supplement and see especially the link to Adolescent Mood Disorders Since 2010: A Collaborative Review.
5. As part of that process, I created a "collaborative review document" with Jean Twenge in 2019. This was a publicly visible Google Doc where we collected all of the studies, surveys, and data sets we could find that shed light on how teen mental health had changed from the early 2000s to the present day in the United States and the U.K. We invited other researchers to add to the doc and critique it. (You can view this collaborative review document and others that I will mention throughout this book at www.anxiousgeneration.com/reviews.)
6. Zahn-Waxler et al. (2008).
7. Askari et al. (2021).
8. The ACHA used only universities that had obtained representative samples using a standardized survey the ACHA had designed. The exact wording of the question was "Within the last 12 months, have you been diagnosed or treated by a professional for any of the following?"
9. American College Health Association (n.d.). You can see the data for female and

male students plotted separately in the online supplement. The patterns are the same, but the rates and increases for women are much higher for anxiety and for depression.

10. Each of the diagnoses in figure 1.2 is increasing, but only the three internalizing disorders are up more than 100%. (The eating disorder anorexia nervosa is related to anxiety and is therefore classified as an internalizing disorder.)

11. The exact wording of the question was "How often did you feel nervous during the past 30 days," and the numbers graphed here are the percent who chose either of the highest two options, out of five: "all of the time" or "most of the time." This question was asked only to those high school seniors who were 18 or older. U.S. National Survey on Drug Use and Health, re-graphed from Goodwin et al., 2020).

12. American Psychiatric Association (2022, p. 215).

13. Parodi et al. (2022). The nationally representative NSDUH survey found similar results with rates for 18–25-year-old girls increasing from 26.13% in 2010 to 40.03% in 2021, while boys increased from 17.35% to 20.26%.

14. The corresponding numbers for depression were 16% "always" or "most of the time," 24% "about half the time," and 60% "less than half the time" or "never."

15. LeDoux (1996) showed that visual information takes two paths through the brain, one of which gets neural signals to the amygdala and hypothalamus nearly immediately, while information in the other path reaches the visual processing areas of the occipital cortex.

16. For a review of anxiety and anxiety disorders, see Wiedemann (2015) and Szuhany & Simon (2022).

17. My description of depression is taken primarily from the chapter on depressive disorders in the *DSM-5-TR*, American Psychiatric Association (2022).

18. Shakespeare, *Hamlet*, 1.2.133–134.

19. Friedman, R. (2018, September 7). The big myth about teenage anxiety. *New York Times*. www.nytimes.com/2018/09/07/opinion/sunday/teenager-anxiety-phones -social-media.html.

20. U.S. Centers for Disease Control, National Center for Injury Prevention and Control. I first encountered a version of this graph in Mercado et al. (2017). I then went to the original source to add on more recent years.

21. You can see graphs of all these trends in the online supplement. The rate for all women over 24 went *down* 25% in that time period.

22. Centers for Disease Control and Prevention. (n.d.).

23. The graph for older teens is quite similar, and can be viewed along with many other graphs in the online supplement.

24. Girls suffer more depression and make more suicide attempts, but they tend to use methods that are reversible, such as cutting their wrists or taking an overdose of sleeping pills. Boys make fewer attempts, but when they do attempt, they are more likely to use methods that can't be reversed, such as a gun or jumping from a tall building.

25. Ortiz-Ospina, E. (2019, September 18). The rise of social media. Our World in Data. ourworldindata.org/rise-of-social-media.

26. I note that the number of liberal democracies around the globe reached its peak during that decade, as I'll discuss in my next book, *Life After Babel*.

27. Chapters 2, 5, and 6 will explain the many mechanisms by which social media damages mental health.

28. Lenhart (2012).

29. Lauricella et al. (2016).

30. Rideout (2021).

31. The report notes, "Much of this frenzy of access is facilitated by mobile devices" (Lenhart, 2015).

32. The largest recipients of teenage attention were five platforms: YouTube, Tik-Tok, Instagram, Snapchat, and Facebook. In fact, 35% of American teens said that they were on at least one of those platforms "almost constantly" (Vogels et al., 2022).

33. Turkle (2015, p. 3).

34. Samsung had introduced smartphones for the Android operating system in 2009.

35. Systrom, K. (2013, February 5). Introducing your Instagram feed on desktop. Instagram. about.instagram.com/blog/announcements/introducing-your-instagram-feed-on-desktop.

36. Protalinski, E. (2012, May 1). Instagram passes 50 million users. *ZDNET*. www.zdnet.com/article/instagram-passes-50-million-users/.

37. Iqbal, M. (2023, May 2). Instagram revenue and usage statistics (2023). *Business of Apps*. www.businessofapps.com/data/instagram-statistics/.

38. The Sandy Hook shooting was one of the most horrific of many mass shootings at American schools. A mentally ill young man charged into an elementary school in Newtown, Connecticut, and murdered 20 children—all aged six or seven—along with six adults.

39. Vermeulen (2021). Also see Twenge (2023, October 24) where she lays out 13 other theories that people have raised as possible explanations for the youth mental health crisis, and why 12 of them don't hold up to scrutiny. Note that Twenge and I both believe that one of those alternative theories is right and important. Alternative #6: "It's because children and teens have less independence."

40. U.S. Bureau of Labor Statistics. (n.d.). Depression data is from the Substance Abuse and Mental Health Services Administration. (2023). *National Survey on Drug Use and Health.*

41. This was one of Durkheim's (1897/1951) findings in his masterpiece *Suicide, a Study in Sociology.* It has been confirmed by later research, for example, Rojce-wicz (1971) and Lester (1993).

42. Bauer et al. (2016).

43. Klar & Kasser (2009). The quotation is from Petré, R. (2010, May 12). Smile, you're an activist! *In These Times*. inthesetimes.com/article/smile-youre-an-activist.

44. Conner, Crawford & Galiotor (2023); Latkin et al. (2022).

45. Belsie, L. (2011). Why Canada didn't have a banking crisis in 2008. National Bureau of Economic Research. www.nber.org/digest/dec11/why-canada-didnt-have-banking-crisis-2008.

46. See my review document, The Coddling of the Canadian Mind? A Collaborative Review, www.anxiousgeneration.com/reviews. See especially Garriguet (2021, p. 9, chart 6).

47. Garriguet (2021). Portrait of youth in Canada: Data report.

48. See the online supplement. Since 2010, suicide rates have been rising for Canadian adolescent girls, but not boys. For boys, this is a pattern I find in many

countries: Depression and anxiety rates tend to go together, while suicide rates are somewhat more variable. For girls, anxiety, depression, self-harm, and suicide rates tend to go together. And among girls, suicide rates have been rising across the five Anglosphere countries. Note that suicide is complex and rare; it is influenced by many factors, such as the prevalence of guns in homes, the difficulty of getting emergency psychiatric care, and the level of social integration (as Émile Durkheim showed). It is by far the most serious mental health outcome, but it is not the most reliable indicator of the overall mental health of the population. See Rausch & Haidt (2023, October 30).

49. See my review document, Adolescent Mood Disorders Since 2010: A Collaborative Review, which you can find linked from the online supplement. It includes dozens of studies on trends in the U.K. and the United States. See especially Cybulski et al. (2021).

50. In the U.K., unlike the United States, the boys' self-harm rate goes up by more than the girls' rate in relative terms, although still much less in absolute terms. I should also note that suicide rates in England and Wales had been dropping overall since the 1980s and remained relatively steady through the early 2000s. Yet against that backdrop of decline, the overall suicide rate has been slowly rising since the 2010s, with a particularly rapid rise among teen boys and girls (along with adult males in their 50s and 60s). Note that the base rates for suicide among adolescents in England and Wales are much lower than in the United States. Once again, the relative increases for teen (15–19) girls are larger than for any other group. See Rausch & Haidt (2023, October 30).

51. I re-graphed the data to put boys and girls in the same figure. You can find the graphs for the other age groups in the online supplement. Cybulski et al. (2021). I thank Lukasz Cybulski for sending me the summary data.

52. Rausch & Haidt (2023, March 29). See also the international review docs that Zach Rausch and I curate for many countries, at www.anxiousgeneration.com /reviews.

53. Australian Institute of Health and Welfare (2022). Although this dataset begins in 2007, other measures of mental health outcomes (e.g., self-reported psychological distress) show no rise in the early 2000s, and increases beginning around 2010. See online supplement for more.

54. For Zach's full analysis of mental health changes in the Nordic countries, see Rausch & Haidt (2023, April 19). High level of distress in figure 1.11 refers to those who reported suffering from at least three psychological ailments at least once a week over the last six months. The ailments were selected from a list of four ailments.

55. There are only a few global surveys that examine adolescent mental health trends over time, with PISA and the Health Behavior in School-Aged Children Study (HBSC) being key sources. The HBSC, initiated in 1983, predominantly covers teens in Europe and North America. Using the HBSC, Cosma et al. (2020) found small declines in teenage mental well-being since 2002. However, this decrease was more pronounced in northern and western Europe and in Canada.

56. Health Behaviour in School-Aged Children (HBSC) (2002–2018). Graphs and data were organized and created by Thomas Potrebny and Zach Rausch.

57. I thank Oliver Hartwich at the New Zealand Initiative, who pointed me to these items.

58. Twenge et al. (2021).
59. Twenge et al. (2021). Data from PISA. Survey data on school alienation was not collected in 2006 and 2009. The PISA data is available for download: Organization for Economic Cooperation and Development (OECD). *PISA survey* [Data sets]. www.oecd.org/pisa/data/.
60. Zach and I have been looking for alternate explanations for a long time. Is there anything else, besides the arrival of smartphones and social media, that could affect teens all around the world at the same time, such as a new chemical released widely around 2012? Or perhaps something that happened in the mid-1990s that affected babies in utero? We consider a few such possibilities in our review document, Alternative Hypotheses to the Adolescent Mental Illness Crisis: A Collaborative Review, www.anxiousgeneration.com/reviews.
61. We are collecting international data, and Zach is writing a series of Substack posts exploring mental health trends around the world. You will find links to these posts with updates about our international findings in the online supplement for this chapter.

Chapter 2: What Children Need to Do in Childhood

1. For evidence of a rapid decline in time spent with friends, see Twenge, Spitzberg & Campbell (2019).
2. Walker et al. (2006).
3. Tanner (1990).
4. There are some documented cases of chimpanzee "culture," where a trick about food gathering or processing is handed down within a community. But these cases are few and far between; cultural learning does not seem to be a major form of chimpanzee learning. See Tomasello (1994, pp. 301–317) for a review.
5. This phrase is often attributed either to the great developmental psychologist Jean Piaget or to Maria Montessori, founder of an educational movement that immerses children in opportunities for free play. As far as I can tell, nobody can find a place where either person wrote that phrase, but it is in keeping with both of their philosophies.
6. See the work of Peter Gray, especially Gray et al. (2023); see also my review document Free Play and Mental Health: A Collaborative Review, www.anxious generation.com/reviews.
7. Gray (2018).
8. Gray (2011, p. 444).
9. Brussoni et al. (2012).
10. Gray (2013).
11. See principle 7. Child Rights International Network. (1959, November 20). *UN declaration on the rights of the child (1959)*. archive.crin.org/en/library/legal-database /un-declaration-rights-child-1959.html.
12. The wording of the item was changed in 2018, so later data is not available. The survey offered five response choices about how often students "get together with friends informally." They ranged from "never" to "almost every day." See further exploration in Twenge, Spitzberg & Campbell (2019).
13. Research note: Throughout this book, I'll present a series of graphs that Zach Rausch and I created from the Monitoring the Future (MTF) survey data (like

figure 2.1). MTF surveys 8th, 10th, and 12th graders about many attitudes and behaviors every year. I will usually show graphs that take the average of all three grades, to give the most comprehensive picture of what is happening to American teens. I will almost always plot the data for boys and girls separately. While MTF began collecting 12th-grade data in 1976, the data gathering for 8th and 10th graders commenced only in 1991, and some variables were later introduced; for instance, weekly social networking usage was added in 2013. At times, I'll show data exclusively from the 12th graders to extend our historical perspective back to the 1970s. I deliberately end most graphs in 2019, even though data is available through 2021, because the COVID pandemic made responses jump around in ways that often distract from the main message about what happened during the Great Rewiring (2010–2015). Also, sample sizes were much smaller in 2020 and 2021, making them less reliable. All graphs show data with the recommended weighting applied and grouped into two-year buckets (for example, data for 2018 and 2019 are averaged together). I do this because plotting every single year often produces a spikiness that obscures underlying trends. Merging pairs of years effectively smooths out the lines to reveal the trends. However, to be complete in my presentation of data, I also show versions of each graph in the online supplement—versions that plot every single year and that run through 2021. For graphs where I showed only 12th graders in the text, I give the graph with the three grades pooled in the supplement if data is available for the younger grades. You can download the MTF data for yourself, along with all other data used this book, at github.com/AfterBabel.

14. Sherman et al. (2009).
15. Cohn & Tronick (1987); Beebe et al. (2010); Wass et al. (2020).
16. Auxier et al. (2020, July 28).
17. National Institute of Play. (n.d.). *Attunement Play.* www.nifplay.org/what-is-play /types-of-play/attunement-play.
18. Ehrenreich (2006); McNeill (1995).
19. Durkheim (1912/1951).
20. Wiltermuth & Heath (2009).
21. See, for example, GlobalWebIndex (2018), which estimated 3 hours per day for ages 16–24 back in 2018. In its 2021 report, GlobalWebIndex found Gen Z using social media platforms 3–4 hours per day in all regions of the world other than Asia-Pacific; Common Sense Media's (2021) census reports lower numbers in its survey of American teens: For those who said that they use social media, boys reported using it an average of 1 hour and 42 minutes per day, while girls reported an average of 2 hours and 22 minutes (Rideout et al., 2022).
22. George & Haidt (2023).
23. Richerson & Boyd (2004). The theory of gene-culture coevolution was developed by Boyd & Richerson (1985); Joe Henrich was a student of Boyd's who developed the theory further.
24. In chapter 5, I will offer a definition of social media. Although streaming platforms like Netflix and Hulu contribute to socialization, social media's unique elements such as social validation, frequent reinforcement for behaviors, public displays of followers and likes, and profiles of relatable peers slightly older than the user make it an even more potent force.
25. Henrich's first paper on prestige bias was written with Francisco Gil-White

(2001). Henrich developed the argument in many later works, including his book *The Secret of Our Success* (2015).

26. Sean Parker in *Axios*: Allen, M. (2017, November 9). Sean Parker unloads on Facebook: "God only knows what it's doing to our children's brains." *Axios*. www.axios.com/2017/12/15/sean-parker-unloads-on-facebook-god-only-knows -what-its-doing-to-our-childrens-brains-1513306792.

27. According to Wikipedia, the phrase was first used by the British journalist Malcolm Muggeridge in 1967, writing that "in the past if someone was famous or notorious, it was for something—as a writer or an actor or a criminal; for some talent or distinction or abomination. Today one is famous for being famous. People who come up to one in the street or in public places to claim recognition nearly always say: 'I've seen you on the telly.'"

28. Black et al. (1998).

29. McAvoy, T. D. (1955). Photograph of Dr. Lorenz studying unlearned habits of ducks and geese at Woodland Institute. Shutterstock.

30. McCabe (2019).

31. On sensitive periods, see Zeanah et al. (2011).

32. Johnson & Newport (1989).

33. Minoura (1992).

34. Minoura (1992, p. 327).

35. Orben et al. (2022). Note that there was also an unexpected appearance of a sensitive period around age 19 for both sexes, but this was thought to relate more to life circumstances, as teens often move away from home at that age, rather than indicating a biological sensitive period.

36. See also a research project from Sapien Labs that surveyed tens of thousands of young adults around the world in 2023. They found that there was a direct linear relationship between the age at which young adults had received their first smartphone and their mental health as an adult: Those whose parents waited longer had better mental health on almost every measure than those who got a phone in elementary or middle school. This study of phone acquisition failed to find a specific sensitive period; rather, it found a cumulative effect of harm throughout childhood (Sapien Labs, 2023).

Chapter 3: Discover Mode and the Need for Risky Play

1. Ingraham, C. (2015, April 14). There's never been a safer time to be a kid in America. *Washington Post*. www.washingtonpost.com/news/wonk/wp/2015/04/14/theres -never-been-a-safer-time-to-be-a-kid-in-america; Let Grow. (2022, December 16). Let Grow takes a look at crime statistics. letgrow.org/crime-statistics/.

2. Bowles, N., & Keller, M. H. (2019, December 7). Video games and online chats are "hunting grounds" for sexual predators. *New York Times*. www.nytimes.com /interactive/2019/12/07/us/video-games-child-sex-abuse.html.

3. Horwitz, J., & Blunt, K. (2023, June 7). Instagram connects vast pedophile network. *Wall Street Journal*. www.wsj.com/articles/instagram-vast-pedophile-network -4ab7189.

4. Richerson & Boyd (2004).

5. The BIS-BAS theory was originally proposed by Gray (1982). For a more recent review see Bijttebier et al. (2009).

6. I take the labels "discover mode" and "defend mode" from Caroline Webb's excellent 2016 book, *How to Have a Good Day*.

7. See, for example, Petersen, A. (2016, October 10). Students flood college mental-health centers. *Wall Street Journal*. www.wsj.com/articles/students-flood-college-mental-health-centers-1476120902.

8. A version of this graph first appeared in *The Wall Street Journal*: Belkin, D. (2018, May 4). Colleges bend the rules for more students, give them extra help. *Wall Street Journal*. www.wsj.com/articles/colleges-bend-the-rules-for-more-students-give-them-extra-help-1527154200. Zach Rausch and I obtained the data from HERI and re-created the graph, adding additional years. Higher Education Research Institute (HERI). (2023).

9. See examples in *The Coddling of the American Mind* (2018), and also see Gosden, E. (2016, April 3). Student accused of violating university "safe space" by raising her hand. *Telegraph*. www.telegraph.co.uk/news/2016/04/03/student-accused-of-violating-university-safe-space-by-raising-he.

10. See my review document, The Coddling of the Canadian Mind? A Collaborative Review, www.anxiousgeneration.com/reviews.

11. Taleb (2012).

12. Gilbert, D. (2004). The surprising science of happiness. TED. www.ted.com/talks/dan_gilbert_the_suprising_science_of_happiness.

13. Phelan (2010).

14. Raudino et al. (2013); Shoebridge & Gowers (2000). For reviews and an updated list, see section 7 of Free Play and Mental Health: A Collaborative Review, www.anxiousgeneration.com/reviews.

15. Sandseter & Kennair (2010). See also their more recent essay: Sandseter et al. (2023).

16. Poulton & Menzies (2002a, 2002b).

17. Sandseter et al. (2023).

18. Used with permission from the collections of the Dallas History & Archives Division, Dallas Public Library.

19. Video games are certainly challenging and exciting, but they do not offer the anti-phobic benefits of risky play (although virtual reality has been found useful as part of exposure therapy in the treatment of specific types of phobias). See Botella et al. (2017).

20. See a collection of photos here: The dangerous playgrounds of the past through vintage photographs, 1880s–1940s. (2023, January 29). Rare Historical Photos. rarehistoricalphotos.com/dangerous-playgrounds-1900s.

21. Kitzman, A. (2023). *Merry go round* [Photograph]. Shutterstock.

22. See research on "adventure playgrounds," described by Rosin, H. (2014, April). The overprotected kid. *Atlantic*. www.theatlantic.com/magazine/archive/2014/04/hey-parents-leave-those-kids-alone/358631/. See Barry, H. (2018, March 10). In Britain's playgrounds, "bringing in risk" to build resilience. *New York Times*. www.nytimes.com/2018/03/10/world/europe/britain-playgrounds-risk.html; Whipple, T. (2019, January 25). Taking risk out of children's lives is putting them in danger. *The Times*. www.thetimes.co.uk/article/taking-risk-out-of-children-s-lives-is-putting-them-in-danger-v7fzcs8b7.

23. Sagdejev, I. (2009). *Hampton forest apartment homes playground* [Photograph]. Wikimedia Commons. commons.wikimedia.org/wiki/File:2009-04-21_Hampton_Forest_Apartment_Homes_playground.jpg.

24. Photo by Jayne Riew.

25. Nauta et al. (2014).

26. See Brussoni's video and project at outsideplay.ca/.

27. Brussoni et al. (2012, p. 3134).

28. Hofferth & Sandberg (2001); Kemple et al. (2016).

29. Tremblay, M. S., & Brussoni, M. (2019, December 16). If in doubt, let them out—children have the right to play. *Conversation*. theconversation.com/if-in-doubt-let-them-out-children-have-the-right-to-play-128780. See also the decline in walking to school (Buliung et al., 2009); Canadian parents and legislators should read the work of Mariana Brussoni at spph.ubc.ca/faculty/mariana-brussoni/.

30. O'Brien & Smith (2002); Dodd et al. (2021); Shaw et al. (2015).

31. Thanks to Eli Finkel, who re-created the graph from the original study (Ramey & Ramey, 2009) in his book *The All-or-Nothing Marriage* and then provided me with the data points to create my own graph.

32. Hofferth & Sandberg (2001).

33. Mullan (2018, 2019).

34. This focus on rising competition and inequality is also the thesis of Doepke et al. (2019).

35. Lareau (2003).

36. DeLoache et al. (2010).

37. Ishizuka (2018).

38. See, for example, Putnam (2000).

39. Gemmel et al. (2023). Also, smaller families meant fewer kids to play with.

40. Furedi (2001). Greg and I had a chapter in *The Coddling of the American Mind* titled "Paranoid Parenting," which was influenced by Furedi, but to our great regret we had failed to quote or cite Furedi directly.

41. See summary in Tiffany, K. (2021, December 9). The great (fake) child-sex-trafficking epidemic. *Atlantic*. www.theatlantic.com/magazine/archive/2022/01/children-sex-trafficking-conspiracy-epidemic/620845/.

42. For an overview of the day care sex abuse panics and the false charges they engendered, see Casey, M. (2015, July 31). How the day care child abuse hysteria of the 1980s became a witch hunt. *Washington Post*. www.washingtonpost.com/opinions/a-modern-witch-hunt/2015/07/31/057effd8-2f1a-11e5-8353-1215475949f4_story.html. See also Day-care sex-abuse hysteria. (2023, June 23). In Wikipedia. Accessed June 28, 2023, en.wikipedia.org/wiki/Day-care_sex-abuse_hysteria.

43. Furedi (2001, p. v).

44. Hillman et al. (1990).

45. Coughlan, S. (2014, December 23). Childhood in the US "safer than in the 1970s." BBC. www.bbc.com/news/education-30578830.

46. For an infuriating recent example, see Skenazy, L. (2022, November 16). Suburban mom handcuffed, jailed for making 8-year-old son walk half a mile home. *Reason*. reason.com/2022/11/16/suburban-mom-jailed-handcuffed-cps-son-walk-home/.

47. For a review of the research indicating that the deprivation of play and autonomy can increase risk of anxiety disorders, see Gray et al. (2023).

48. Haslam (2016).

49. See the online supplement for the Ngram graph of the term "emotional safety."

50. Edmondson (1999).

51. Haefeli, W. (2004) We've Created a Safe poster. *The New Yorker* © Condé Nast.

52. Lukianoff and Haidt (2018, p. 27). We thank Pamela Paresky for inventing the term.

53. Photo by Robert Strand.

54. See Pew Research Center (2015, pp. 50–51). The ages parents give are influenced by how safe they perceive their neighborhoods to be, but not by much. Parents who say that their neighborhood is an excellent or very good place to raise kids offer ages just one year less than those I listed in the text. See also for similar results Grose, J., & Rosin, H. (2014, August 6). The shortening leash. *Slate*. www
.slate.com/articles/life/family/2014/08/slate_childhood_survey_results_kids
_today_have_a_lot_less_freedom_than_their.html.

55. Fay, D. (2013). Diagram of a secure attachment [Photograph]. In *Becoming safely attached: An exploration for professionals in embodied attachment*. dfay.com
/archives/3134. Box and right-hand text added by Haidt.

56. See chapter 7 of Ames and Ilg's 1979 book, *Your Six-Year-Old: Loving and Defiant*, which lists the things a six-year-old should be able to do by the start of first grade, including "Can he travel alone in the neighborhood (4–8 blocks) to store, school, playground, or to a friend's home?"

Chapter 4: Puberty and the Blocked Transition to Adulthood

1. Hebb (1949).

2. The cement analogy overstates the lockdown. The brain continues to be malleable for life in the sense of forming new synapses, and there are some areas of the brain where new neurons continue to grow during adulthood. Adults continue to learn, and all learning involves some kind of brain change. But structural change is far more limited after areas have transformed during puberty.

3. Steinberg (2023); Fuhrmann et al. (2015).

4. Steinberg (2023, p. 26).

5. See, for example, Hara Marano's 2008 book, *A Nation of Wimps*. For evidence that the millennials developed an increasingly external locus of control, see Twenge et al. (2004).

6. Although Twenge (2023b) showed that there was a smaller rise among millennials, which began a year or two *after* the surge among Gen Z. Furthermore, Gray et al. (2023) have argued that children's independence has been decreasing since the 1940s, and that some measures of psychopathology have been increasing, slowly, since then too. I acknowledge this point, this backstory, but because rates generally flat or even improving in the 1990s and early 2000s, I am focused on why there was a surge of mental illness that began in the early 2010s.

7. The ability to process rapidly flashing images and written messages while multitasking on multiple screens was just not a selection pressure in human evolution, so it is not a skill that needs to be practiced in childhood. Even if today's children will need to do that as adults, immersing children in such stimulation early does not make them better prepared for the future.

8. Brown (1991).

9. For my account of the sunrise dance I draw on Markstrom (2010) and Marks

(1999). Apache female puberty sunrise ceremony. Web Winds. www.webwinds .com/yupanqui/apachesunrise.htm.

10. Lacey (2006).

11. I obtained this clarification from Uri Bilmes, who had been trained as a rabbi: "It's important to note that the age thresholds for adulthood were established in a different time and society. For reference, a famous passage in the rabbinic corpus lists different ages and their ideal corresponding stages of development in the following way: 'Age five for Bible study, age 10 for Mishna study, age 13 for obligation in commandments, age 15 for Talmudic study, age 18 for marriage...' In the world of 13-year-old adulthood, marriage was not to be pushed off beyond 18. In today's society, the consideration of a 7th grader as a 'man' (whose mom still packs his lunch) is almost an anachronism, maybe even humorous."

12. Markstrom (2011, p. 157).

13. For street gang initiation rites, see Descormiers and Corrado (2016).

14. Nuwer (1999); Kim, Y. (2018, July 10). 8 girls get real about their crazy sorority experiences. *Seventeen*. www.seventeen.com/life/real-girl-stories/a22090500 /craziest-sorority-hazing-stories/.

15. This is found in the Monitoring the Future data that you can see in the online supplement. Also see Burge (2021).

16. Of course the lines at 13 and 18 were blurry in practice. Kids could get into movies before 13 and into bars before 18 if they had a fake ID. But there was some risk involved—some actual fear when you handed your fake ID to the bartender or bouncer.

17. Three items come from the Monitoring the Future data set: drinking alcohol: "Have you ever had beer, wine, or liquor to drink?"; work: "On the average over the school year, how many hours per week do you work in a paid or unpaid job?"; driver's license: "Do you have a driver's license?" The final item, sex, comes from the CDC Youth Risk Behavior Survey: "Have you ever had sexual intercourse?"

18. Rideout et al. (2022) reports that 18% of 8–12-year-olds are now daily users of social media, mostly Snapchat and Instagram. If we could limit this finding to 11- and 12-year-olds, it would be much higher.

19. As Ron Lieber says in his excellent 2015 book, *The Opposite of Spoiled*, "Every conversation about money is also about values. Allowance is also about patience... work is about perseverance." He also recommends that a weekly allowance begin "by first grade, at the latest," which is roughly age 6.

20. Personally I think the age for the first smartphone should be 16, but given where we are now, and given the importance of getting smartphones and social media entirely out of the lives of middle school students, I'm proposing that we make the transition to high school—around age 14—be the bright line that anchors a new norm.

Chapter 5: The Four Foundational Harms: Social Deprivation, Sleep Deprivation, Attention Fragmentation, and Addiction

1. Thorndike (1898).

2. John Schroter. (2021, October 8). *Steve Jobs introduces iPhone in 2007* [Video]. YouTube. www.youtube.com/watch?v=MnrJzXM7a6o (time code 2:14); Jobs' original vision for the iPhone: No third-party native apps. (2011, October 21).

9to5Mac. 9to5mac.com/2011/10/21/jobs-original-vision-for-the-iphone-no-third
-party-native-apps.

3. Silver, S. (2018, July 10). The revolution Steve Jobs resisted: Apple's App Store marks 10 years of third-party innovation. *AppleInsider*. appleinsider.com/articles/18/07/10/the-revolution-steve-jobs-resisted-apples-app-store-marks-10-years-of-third-party-innovation.

4. Turner, A. (2023). How many apps in Google Play Store? (August 2023). *BankMy Cell*. www.bankmycell.com/blog/number-of-google-play-store-apps.

5. To grasp the magnitude of the ad-centric model, consider that in 2019, 3.3 billion people used social media on a mobile device. And in 2019 alone, advertising accounted for *98%* of Meta's revenue, which was more than $69 billion. The same advertising-based business model powers TikTok, Snap, and most other major social media platforms. Their enormous revenues are a function of how well they cater to their customers—the advertisers—not the 3+ billion users. See Kemp (2019). That number increased to 4.9 billion in 2023; see Wong & Bottorff (2023).

6. Lenhart (2015).

7. For the history of definitions of social media since 1994, see Aichner et al. (2021).

8. Brady et al. (2017).

9. Pew Research Center (2021).

10. See Halldorsdottir et al. (2021), Verduyn et al. (2015), and Kim et al. (2020), for evidence on the negative mental health effects of passive social media use.

11. The numbers I report for total screen time come from Rideout & Robb (2019): around 5 hours a day of nonschool screen media for kids ages 8–12; 7–8 hours a day for older teens. Nagata, Ganson, et al. (2022) report findings consistent with those numbers: Children ages 9–10 were doing 4 hours of screen time a day before COVID. Nagata, Cortez, et al. (2022) report that 13-year-olds in the ABCD study were spending closer to 8 hours per day on screens in 2021. And the American College of Pediatricians (2020) found similar numbers: about 5 hours per day for 8–12-year-olds; nearly 7.5 hours for teens. All of these studies exclude screen use for school or homework; all are just leisure use, so I report this as around 40 hours per week for tweens and more than 50 hours per week for teens. Similar numbers are found in the U.K.: Hiley, C. (2022, September 13). Screen time report 2022. *Uswitch*. www.uswitch.com/mobiles/screentime-report.

12. Twenge, Martin & Spitzberg (2019), analyzing data from Monitoring the Future.

13. There is less data on technology use trends among Asian Americans. The results are contradictory, with some studies reporting lower screen time compared with white, Black, and Latino teens (see Nagata, Ganson, et al., 2022; Nagata et al. 2023), while others show that screen time use is comparable to Black and Hispanic teens (see Rideout et al., 2011).

14. Research note: In previous decades, the digital divide has resulted in socioeconomic disparities, with wealthier families having quicker adoption and greater access to technology such as computers, laptops, and televisions. Although the digital divide still exists, it often plays out in unexpected ways in the United States. For example, despite 57% of U.S. adults with an income less than $30,000 having high-speed broadband, compared with 83% of those earning between $30,000 and $100,000, lower-income families increasingly rely on smartphones for internet, leading to heavier smartphone usage. Notably, smartphone owner-

ship among tweens (8-to-12-year-olds) and teens (13-to-18-year-olds) doesn't vary significantly across social classes, though time spent on screens does. Tweens from lower-income families (who make less than $35,000 a year) spend about three hours more per day on screens than their wealthier peers, while lower-income teens spend about two hours more. Moreover, many tech executives, including those from Silicon Valley, send their children to private schools like the Waldorf School of the Peninsula where screen use is prohibited. This is in contrast with many public schools that are advancing 1:1 technology programs, trying to give every child their own device. Additionally, many lower-income parents work multiple jobs, and they are more likely to raise a child in a single-parent household, leaving less time and energy to invest in monitoring the quantity and content of their child's screen use. The socioeconomic variation in screen use has been found in other countries too; see, for example, Pedersen (2022) in Denmark. Regarding race, Black and Hispanic youths are more likely to own a smartphone than their white counterparts. Black tweens spend about two more hours per day on screen time than white tweens do. Latino tweens have an even larger gap, spending about two and a half hours more than white tweens do. LGBTQ teens also report spending about three hours more time per day on screens than their cis heterosexual counterparts. For sources, see Vogels (2021); Rideout et al. (2022); Atske and Perrin (2021); Rideout and Robb (2019); Nagata et al. (2023); Assari (2020); Pulkki-Råback et al. (2022); Bowles, N. (2018, October 16). The digital gap between rich and poor kids is not what we expected. *New York Times*. www.nytimes.com/2018/10/26/style/digital-divide-screens-schools.html.

15. Vogels et al. (2022): "Across these five platforms, 35% of all U.S. teens say they are on at least one of them almost constantly."
16. Thoreau (1910, p. 39).
17. Gray (2023).
18. Kannan & Veazie (2023).
19. American Time Use Survey. I thank Dr. Viji Kannan for sending me the data points in Kannan & Veazie (2023), which Zach and I re-graphed to create this figure.
20. Twenge (2017, Chapter 3). See also Twenge, Spitzberg & Campbell (2019). I'll show in chapter 6 that these are not just correlations; experiments show causation, especially for social media.
21. Barrick et al. (2022).
22. Przybylski & Weinstein (2012). For a review of research, see Garrido et al. (2021).
23. *Highlights* (2014, October 14). National survey reveals 62% of kids think parents are too distracted to listen. PR Newswire. www.prnewswire.com/news-releases/national-survey-reveals-62-of-kids-think-parents-are-too-distracted-to-listen-278525821.html.
24. Pew Research Center (2020).
25. With thanks to Jacob Silliker for sharing his insights with me, and for his permission to reprint them here.
26. Hummer & Lee (2016).
27. Tarokh et al. (2016); Lowe et al. (2017).
28. Wolfson & Carskadon (2003); Perez-Lloret et al. (2013).
29. Dahl (2008); Wheaton et al. (2016).

30. Owens et al. (2014); Garbarino et al. (2021).
31. Paruthi et al. (2016).
32. James Maas, quoted in Carpenter, S. (2001, October). Sleep deprivation may be undermining teen health. *Monitor on Psychology, 32*. www.apa.org/monitor/oct01 /sleepteen.
33. National Addiction & HIV Data Archive Program (n.d.-a, n.d.-b). *Monitoring the Future.*
34. Alonzo et al. (2021).
35. Perrault et al. (2019). Also see Garrison & Christakis (2012) and Mindell et al. (2016).
36. For video games, see Peracchia & Curcio (2018). For e-readers, see Chang et al. (2014). For computers, see Green et al. (2017). For social media, see Rasmussen et al. (2020). There are a few studies that report little to no effect of screen use on sleep. See Przybylski (2019).
37. Hisler et al. (2020).
38. There are many studies on this topic. To add some international evidence: a large study (Khan et al., 2023) analyzed the results of a survey given to teens in 38 countries and found that heavy users of all digital media had more sleep problems than did light users, with effects kicking in above two hours a day for each media type, and accelerating above four hours per day (again suggesting that addiction contributes to these effects). The effects were generally larger for girls. I should note that the effects for "passive screen time," which was mostly TV and videos, did not appear until daily average time was above four hours. This is a consistent finding: TV, which is passive, is not as bad as social media or video games, which involve rapid behaviors that get reinforced with rewards, and hence are more addictive.
39. Guo et al. (2022); Ahmed et al. (2022); Kristensen et al. (2021); Alimoradi et al. (2019).
40. As quoted in Hern, A. (2017, April 18). Netflix's biggest competitor? Sleep. *Guardian.* www.theguardian.com/technology/2017/apr/18/netflix-competitor-sleep-uber -facebook.
41. Goldstone et al. (2020).
42. Statista. (2023, April 18). *Weekly notifications from social apps to U.S. Gen Z mobile users 2023.* www.statista.com/statistics/1245420/us-notifications-to-social-app -ios-users. I note that most teens don't use all 13 apps, although the average teen has an account on 7–8 social media platforms; Kemp, S. (2023, January 26). DataReportal. datareportal.com/reports/digital-2023-deep-dive-time-spent-on -social-media. Of course many teens learn to turn notifications off for some apps, and many use built-in features to turn off all notifications temporarily. But my students agree: Their phones interrupt them continuously throughout the day.
43. James (1890, chapter 11).
44. Carr (2012, p. 7).
45. I developed this argument about the necessity of phone-free schools, in Haidt, J. (2023, June 6). Get phones out of school now. *Atlantic.* www.theatlantic.com /ideas/archive/2023/06/ban-smartphones-phone-free-schools-social-media /674304.

46. Kim et al. (2019).

47. Madore & Wagner (2019).

48. Ward et al. (2017). I note that one attempt to replicate the study did not find that phone location affected performance (Ruiz Pardo & Minda, 2022). But other studies have found that when phones are visible, they have disruptive effects. See Dwyer et al. (2018); Tanil & Young (2020); Skowronek et al. (2023).

49. For additional sources on the relationship between ADHD and screen time, see Boer et al. (2019); Liu et al. (2023); Santos et al. (2022); Tamana et al. (2019).

50. Boer et al. (2020).

51. Baumgartner et al. (2018).

52. There is a correlation of heavy or problematic social media use with lower executive function; see Reed (2023). But it is difficult to test experimentally whether long-term use has damaging effects because it would be unethical to randomly assign young people to a heavy social media use condition.

53. See Alavi et al. (2012) & Grant et al. (2010) for discussions around the classification, similarities, and differences between behavioral and chemical addictions.

54. See, for example, Braun, A. (2018, November 13). Compulsion loops and dopamine hits: How games are designed to be addictive. *Make Tech Easier*. www .maketecheasier.com/why-games-are-designed-addictive.

55. I thank Nir Eyal for permission to reprint this figure. Eyal also published a book in 2019 titled *Indistractable: How to Control Your Attention and Choose Your Life*, which provided strategies for breaking bad tech habits.

56. Spence et al. v. Meta Platforms Inc., No. 3:22-cv-03294, N.D. Cal. (San Francisco, 2022), Document 1, pp. 24–25, para. 32. socialmediavictims.org/wp-content /uploads/2022/06/Spence-Complaint-6_6_22.pdf.

57. Lembke (2021, p. 57).

58. American Psychiatric Association (2023, January). See also Marcelline, M. (2022, December 12). Canada judge authorizes *Fortnite* addiction lawsuit. *PCMag*. www.pcmag.com/news/canada-judge-authorizes-fortnite-addiction-lawsuit.

59. Chang et al. (2014).

60. Lembke (2021, p. 1).

61. See especially Maza et al. (2023).

62. U.S. Department of Health and Human Services (2023).

63. Vogels & Gelles-Watnick (2023).

64. Nesi et al. (2023).

65. Berger et al. (2022); Berger et al. (2021); Nagata et al. (2023).

66. See Social Media and Mental Health: A Collaborative Review, which Zach Rausch, Jean Twenge, and I curate. Very few studies find benefits. Available at www .anxiousgeneration.com/reviews.

67. YouTube is technically a form of social media, but it is mostly used as a source of information. It is implicated in radicalization and many other social and psychological problems, but when people rate the pluses and minuses of platforms, YouTube gets among the most positive ratings; see, for example, Royal Society for Public Health (2017).

68. Adding a further reason to doubt: Many of the studies used to support claims about the social and educational benefits of social media are actually reporting results from studies on internet use, and some were conducted before 2012,

before Instagram, Snapchat, and TikTok became popularized. See Uhls et al. (2017) for a review of the benefits. They refer to many studies before 2012 and include sources that focus on internet use, such as Borca et al. (2015).

69. Nesi et al. (2023).
70. Vogels (2022).

Chapter 6: Why Social Media Harms Girls More Than Boys

1. Spence et al. v. Meta Platforms Inc., No. 3:22-cv-03294, N.D. Cal. (San Francisco, 2022), Document 1, pp. 110–111, para. 187. socialmediavictims.org/wp-content/uploads/2022/06/Spence-Complaint-6_6_22.pdf. Drawing used with permission from Alexis's parents. I am working with the law firm representing the Spences.

2. Several studies have found links between social media use and suicidal thoughts, for girls but not for boys. See Coyne et al. (2021). See also Brailovskaia, Krasav-tseva, et al. (2022), who looked only at women in Russia. They found that "problematic social media use significantly mediated the relationship between daily stress and suicide-related outcomes," but this was true only for younger women (below age 29), not for older women.

3. See Rausch & Haidt (2023, March 29).

4. Research note: The graphs in this chapter are mostly about American teens because there is excellent data on them going back in some studies to the 1970s, particularly the Monitoring the Future study. I am confident that these trends are similar in the other Anglosphere countries. I believe—from some large international studies and from what people write to me—that these trends are happening in much of Europe and Latin America. I know little about trends in Asia or Africa, although the isolating, loneliness-inducing effects of rapid technological change on social relationships may be muted in more collectivist, religious, and family-oriented societies. Sources: Rausch (2023, March); also the international reviews that Zach Rausch and I curate, available at www.anxiousgeneration.com/reviews.

5. Orben & Przybylski (2019).

6. Twenge, Haidt, et al. (2022). We reanalyzed the same data sets used by Orben & Przybylski (2019), and we addressed a few other problems we saw in that study, such as controlling for psychological variables that are related to mental health, rather than just controlling for demographic variables, as is usually done. We found correlations between social media use and mental health equivalent to $r = 0.20$, which is in the ballpark of binge drinking, not eating potatoes, in those data sets.

7. There has been a surprising convergence in recent years around the size of the correlation between social media use and internalizing disorders (especially anxiety and depression). Jean Twenge and I have found it to be around $r = 0.20$ when you limit the analysis to girls and social media (where r is the "Pearson correlation coefficient" that runs from $r = -1.0$ for a perfect negative correlation, through $r = 0$ for a complete absence of correlation, to $r = 1.0$ for a perfect positive correlation). Orben & Przybylski (2019) said that the correlation was equivalent to $r < 0.04$, which truly would be negligible, but again, that was for all digital activities and all teens. When Amy Orben (2020) reviewed many other studies that were limited to social media (rather than all digital media), she

found that the associations with well-being range from $r = 0.10$ to $r = 0.15$, and that was for boys and girls combined. The effects are usually larger for girls, so that puts it up above $r = 0.15$ for the link between social media and poor mental health for girls, which is very close to what Twenge and I have found. Jeff Hancock, another major researcher who has been skeptical of the claim that social media harms teen mental health, conducted a meta-analysis of studies through 2018 (Hancock et al., 2022). He and his coauthors found that time on social media was not substantially associated with most well-being variables, with the exception of depression and anxiety. For those outcomes the correlations were, once again, between $r = 0.10$ and $r = 0.15$, and that, again, is for boys and girls merged. So the research community is closing in on a consensus that crude measures of social media use are correlated with crude measures of anxiety and depression, for girls, at a level around or above $r = 0.15$. (If measurement of the two variables was better, the correlations would be higher.) Is $r = 0.15$ tiny? Not in public health (see Götz et al., 2022).

8. I note that in Twenge, Haidt, et al (2022) and other studies, "internet use" often shows similarly high correlations with poor mental health, for girls in particular. I also note that some studies find moderator variables—that is, variables that make some girls more or less prone to harm from social media. Some that have been found include early puberty, high media consumption, and preexisting depression or anxiety. See section 2 of Social Media and Mental Health: A Collaborative Review.

9. See a review of these studies in my Substack post: Social Media Is a Major Cause of the Mental Illness Epidemic in Teen Girls. Here's the Evidence (Haidt, 2023, February 23).

10. Denworth, L. (2019, November 1). Social media has not destroyed a generation. *Scientific American*. www.scientificamerican.com/article/social-media-has-not-destroyed-a-generation.

11. Millenium Cohort Study. Analyzed by Kelly et al. (2018). Replotted by Zach Rausch.

12. Some studies do find that adolescents with preexisting depression are more likely to seek out social media. But many studies establish that heavy social media use causes depression, and some of the longitudinal studies establish that increasing social media use at Time 1 predicts greater depression at Time 2. See, for example, Primack et al. (2020); Shakya & Christakis (2017).

13. Hunt et al. (2018, p. 751).

14. Kleemans et al. (2018).

15. In the Social Media and Mental Health collaborative review document that Jean Twenge, Zach Rausch, and I curate, we collect the abstracts of hundreds of studies on social media, organized by whether they find evidence of harm or not. As I write, in 2023, we have 20 RCT studies in the document, of which 14 (70%) found evidence of harm. Of the 6 experiments that did not find evidence of harm, it is noteworthy that 4 of them asked participants to give up social media for a short time period—one week or less. I believe we should not expect benefits from making addicts give up their drug for a short time. They need at least three weeks for their brains to reset and get past withdrawal cravings. Of the 14 that found an effect, only 2 used a time interval of one week or less. So if we eliminate the 6 studies that used short durations, we are left with a ratio of 12 to 2, or 86% that found a significant effect.

16. This is known as *Metcalfe's law*: The financial value or influence of a telecommunications network is proportional to the *square* of the number of connected users of the system. Metcalfe's law (2023, June 27). Wikipedia. Accessed July 10, 2023, en.wikipedia.org/wiki/Metcalfe%27s_law.

17. This is what teachers tell me, and it is what I see even among MBA students at NYU Stern. An increase in conversation and laughter is widely reported as an effect when schools require phones to be locked away during the school day; see Cook, H. (2018, February 20). Noise levels dialed up as school's total phone ban gets kids talking. *Age*. www.theage.com.au/national/victoria/noise-levels-dialled-up-as-school-s-total-phone-ban-gets-kids-talking-20180220-p4z0zq.html.

18. See Twenge, Spitzberg & Campbell (2019) for evidence and elaboration on this point.

19. These studies are sometimes called quasi-experiments because the researchers take advantage of natural variation in the world as though it were random assignment. You can find these studies in section 4 of the collaborative review document Social Media and Mental Health: A Collaborative Review, www.anxiousgeneration.com/reviews.

20. Braghieri et al. (2022, p. 3660). For a critique of this study see Stein (2023). I believe the basic "difference-in-differences" design is sound; it gets at the relevant comparison of entire colleges where most people adopted Facebook at the same time, versus colleges where adoption was slower.

21. Arenas-Arroyo et al. (2022, p. 3). The study found particular damage to father-daughter relationships, although the effect was limited to those relationships that were already strained.

22. See Social Media and Mental Health: A Collaborative Review, www.anxiousgeneration.com/reviews.

23. I note that several prominent researchers disagree with me on these points. They don't assert that social media is harmless, but they believe that the accumulated scientific evidence is not yet sufficient to prove that social media causes anxiety, depression, and other negative psychological outcomes. You can find links to these researchers' objections, along with my responses, on my Substack. See my post: Why Some Researchers Think I'm Wrong About Social Media and Mental Illness (Haidt, 2023, April 17).

24. Lenhart (2015).

25. Royal Society for Public Health (2017).

26. Research note: It is very difficult for any of us to accurately respond to these time estimation questions, and some researchers have questioned the utility of such self-report data; see Sewall et al. (2020). But the pattern of rising super-use is validated by the increase that Pew finds in the percentage of U.S. teens who say that they are online "almost constantly" (Perrin & Atske, 2021).

27. A 2023 Common Sense report indicates that among 11-to-15-year-old girls who actively use these platforms, the average daily usage is as follows: TikTok at 2 hours and 39 minutes, YouTube at 2 hours and 23 minutes, Snapchat at 2 hours, and Instagram at 1 hour and 32 minutes. See Nesi et al. (2023).

28. The question used in the survey in 2013 and 2015 was "About how many hours a week do you spend visiting social networking websites like Facebook?" In 2017, the item changed to "About how many hours a week do you spend visiting social networking websites like Facebook, Twitter, Instagram, etc."

29. Chen et al. (2019). Also see Eagly et al. (2020), who analyzed U.S. public opinion

polls from 1946 through 2018 and found that over these years people increasingly saw women as more affectionate and emotional (communion traits), while views of men as ambitious and courageous (agency traits) remained the same.

30. Guisinger & Blatt (1994).

31. Hsu et al. (2021).

32. See Maccoby & Jacklin (1974); Tannen (1990) for a review of gender differences in language use; Todd et al. (2017).

33. Kahlenberg & Wrangham (2010); Hassett et al. (2008).

34. "Jealousy, Jealousy" by Olivia Rodrigo is available on YouTube, just search for those words.

35. Fiske (2011, p. 13).

36. Leary (2005).

37. I thank @JosephineLivin for creating this image and for giving me permission to use it.

38. Josephs, M. (2022, January 26). 7 teens on Instagram filters, social media, and mental health. *Teen Vogue.* www.teenvogue.com/story/7-teens-on-instagram -filters-social-media-and-mental-health

39. Curran & Hill (2019) analyzed studies of perfectionism from the United States, the U.K., and Canada since 1989. They found that self-oriented (SOP), other-oriented (OOP), and socially prescribed perfectionism (SPP) had been rising linearly during this period, with no bend or acceleration in the trend line. However, Zach and I noticed that the data points for socially prescribed perfectionism, upon which the trend line was based, seemed to have a bend in the curve, with a sharp upturn around 2010. We contacted the authors about this, and Dr. Curran stated, "You're right to point out that the trend in our 2017 paper appears quadratic. Indeed, I reanalysed the data with the most up-to-date SPP scores for my book and ran the quadratic model, which was a better fit than the linear one." You can view the updated quadratic figure, with the upward bend in 2010, in the online supplement.

40. Torres, J. (2019, January 13). How being a social media influencer has impacted my mental health. *HipLatina.* hiplatina.com/being-a-social-media-influencer -has-impacted-my-mental-health.

41. Chatard et al. (2017). See also Joiner et al. (2023), who found that young women who watched thin women do TikTok dances felt worse about their own bodies, while young women who watched heavy women do TikTok dances felt better about their own bodies.

42. [iamveronika]. (2021, August 10). Suicidal because of my looks [Online forum post]. Reddit. www.reddit.com/r/offmychest/comments/p22en4/suicidal_because_of _my_looks.

43. Hobbs, T. D., Barry, R., & Koh, Y. (2021, December 17). "The corpse bride diet": How TikTok inundates teens with eating-disorder videos. *Wall Street Journal.* www.wsj.com/articles/how-tiktok-inundates-teens-with-eating-disorder-videos -11639754848.

44. Wells, G., Horwitz, J., & Seetharaman, D. (2021, September 14). Facebook knows Instagram is toxic for teen girls, company documents show. *Wall Street Journal.* www.wsj.com/articles/facebook-knows-instagram-is-toxic-for-teen-girls -company-documents-show-11631620739.

45. Archer (2004).

46. Crick & Grotpeter (1995); Archer (2004).

47. Kennedy (2021).

48. Girls who reported having been cyberbullied in the past 12 months increased from 17% in 2006, to 27% in 2012. Schneider et al. (2015).

49. Li et al. (2020, Table 2).

50. Lorenz, T. (2018, October 10). Teens are being bullied "constantly" on Instagram. *Atlantic*. www.theatlantic.com/technology/archive/2018/10/teens-face -relentless-bullying-instagram/572164.

51. India, F. (2022, July 22). Social media's not just making girls depressed, it's making us bitchy too. *New Statesman*. www.newstatesman.com/quickfire/2022/07/social -media-making-young-girls-depressed-bitchy.

52. See the case of Molly Russell, in the U.K., whose suicide was found to be caused in large part by bullying on social media platforms. Also see this article for a young person's review of the effect of these platforms: Gevertz, J. (2019, February 10). Social media was my escape as a teenager—now it's morphed into something terrifying. *Independent*. www.independent.co.uk/voices/facebook-twitter -young-people-mental-health-suicide-molly-russell-a8772096.html.

53. See Fowler & Christakis (2008).

54. See Rosenquist et al. (2011).

55. Tierney & Baumeister (2019).

56. Boss (1997). She used the term "epidemic hysteria." I have swapped in the term "sociogenic illness" because it is more descriptively accurate in pointing to social causes, because it has been used by researchers more recently, and because the term "hysteria" has often been used in ways that denigrate women.

57. Waller (2008).

58. See Wessely (1987) for an academic account, and for a journalistic account of the two variants and the sex difference usually observed, see Morley, C. (2015, March 29). Carol Morley: "Mass hysteria is a powerful group activity." *Guardian*. www.theguardian.com/film/2015/mar/29/carol-morley-the-falling-mass -hysteria-is-a-powerful-group-activity.

59. For a sad example, see Gurwinder's profile of Nicholas Perry, a young man who was trained by audience capture to eat to the point of extreme obesity: Gurwinder. (2022, June 30). The perils of audience capture. *The Prism*. gurwinder .substack.com/p/the-perils-of-audience-capture.

60. Jargon, J. (2023, May 13). TikTok feeds teens a diet of darkness. *Wall Street Journal*. www.wsj.com/articles/tiktok-feeds-teens-a-diet-of-darkness-8f350507.

61. Müller-Vahl et al. (2022).

62. For a journalistic account of these cases, see Browne, G. (2021, January 9). They saw a YouTube video. Then they got Tourette's. *Wired*. www.wired.co.uk/article /tourettes-youtube-jan-zimmermann.

63. You can watch her TikTok videos here: Field, E. M. [@thistrippyhippie]. (n.d.). [TikTok profile]. TikTok. www.tiktok.com/@thistrippyhippie?lang=en.

64. The *DSM-5* estimates that the 12-month prevalence of DID among U.S. adults is 1.5% (American Psychiatric Association, 2022, March). However, the population estimates are still debated, with studies showing variation, though generally falling between 1% to 1.5% of the U.S. population. See Dorahy et al. (2014); Mitra & Jain (2023). Part of the reason for the range (which is sometimes reported as being higher than 1.5%) is that psychiatrists have long debated whether or not

it is a real disorder. Some believe that it is a form of post-traumatic stress disorder, a reaction to trauma so severe that the mind forms multiple identities to cope. Others believe that the emergence of DID relies heavily on suggestion and a predisposition to fantasy and suggestibility, which may occur in the wake of real trauma. For a discussion of "myths" around DID, see Brand et al. (2016).

65. Rettew, D. (2022, March 17). The TikTok-inspired surge of dissociative identity disorder. *Psychology Today*. www.psychologytoday.com/gb/blog/abcs-child -psychiatry/202203/the-tiktok-inspired-surge-dissociative-identity-disorder.

66. Lucas, J. (2021, July 6). Inside TikTok's booming dissociative identity disorder community. *Inverse*. www.inverse.com/input/culture/dissociative-identity-disorder -did-tiktok-influencers-multiple-personalities.

67. Styx, L. (2022, January 27). Dissociative identity disorder on TikTok: Why more teens are self-diagnosing with DID because of social media. *Teen Vogue*. www .teenvogue.com/story/dissociative-identity-disorder-on-tiktok.

68. American Psychiatric Association (2022, pp. 515, 518); for an estimate of 1% of the youth population in the United States, see Turban & Ehrensaft (2018).

69. Block (2023); Kauffman (2022); Thompson et al. (2022). Turban et al. (2022) note that using YRBS data, they saw a decline in the number of youth identifying as transgender and gender diverse from 2017 to 2019.

70. Aitken et al. (2015); de Graaf et al. (2018); Wagner et al. (2021); Zucker (2017). However, some researchers argue that the gap has not reversed and the ratio of natal males to natal females is now 1.2 to 1; see Turban et al. (2022).

71. Haltigan et al. (2023); Littman (2018); Marchiano (2017).

72. Coleman et al. (2022); Littman (2018); Littman (2021);

73. Coleman et al. (2022); Kaltiala-Heino et al. (2015); Zucker (2019).

74. See Buss's 2021 book, *When Men Behave Badly*. Each chapter explores elements of male psychology that seem to have been adaptive for some extended period of human evolution—a time when most males never got a chance to mate, so competition among males was intense, and violence sometimes "paid," in evolutionary terms, if it led to even a single act of copulation. Buss says repeatedly that the evolutionary frame in no way condones sexual aggression or implies that change is impossible. Instead, evolutionary psychology can help us understand why sexual aggression is far more common among males and how we can effectively reduce it.

75. Culture and socialization can discourage the use of such tactics and shame men who use them; indeed, the feminist movement from the 1970s through #MeToo has brought about such changes. And yet, as society fragments into millions of online communities, some of those communities put men into competition for prestige by espousing more and more extreme attitudes, making those tactics seem permissible again.

76. See Mendez, M., II. (2022, June 6). The teens slipping through the cracks on dating apps. *Atlantic*. www.theatlantic.com/family/archive/2022/06/teens-minors-using -dating-apps-grindr/661187.

77. See Thorn & Benenson Strategy Group (2021); Bowles, N., & Keller, M. H. (2019, December 7). Video games and online chats are "hunting grounds" for sexual predators. *New York Times*. www.nytimes.com/interactive/2019/12/07/us/video -games-child-sex-abuse.html.

78. Sales (2016, p. 110).

79. Sales (2016, pp. 49–50).
80. Sales (2016, p. 216).
81. deBoer, F. (2023, March 7). Some Reasons Why Smartphones Might Make Adolescents Anxious and Depressed. *Freddie deBoer*. https://freddiedeboer.substack .com/p/some-reasons-why-smartphones-might
82. Damour (2016).
83. The pooled average for 8th, 10th, and 12th graders shows a similar pattern to the 12th graders. The pooled data begins only in 1997. See the online supplement.

Chapter 7: What Is Happening to Boys?

1. Hari (2022, p. 4).
2. National Addiction & HIV Data Archive Program. (n.d.-a). *Monitoring the Future*.
3. See discussion on suicide in chapter 1, and in Rausch & Haidt (2023, October 30).
4. Zach Rausch is essentially a coauthor with me on this chapter. He has been maintaining a collaborative review document collecting research on boys, and has created a detailed timeline of how technology changed since the 1970s in ways that drew boys in. See the online supplement for links to both documents. We worked out the story told in this chapter together.
5. The American Institute for Boys and Men.
6. One robust difference is the "things versus people" dimension, with men higher than women on interest in things, women higher than men on interest in people (Su et al., 2009).
7. This quote comes from her TED Talk about the book: Rosin, H. (2010, December). New data on the rise of women [Video]. TED. www.ted.com/talks/hanna_rosin _new_data_on_the_rise_of_women/transcript.
8. Rosin (2012, p. 4).
9. See Parker (2021). The same is found for postgraduate degrees (Statista Research Department, 2023). I note that this chapter draws mostly on the voluminous statistics available for the United States, but Reeves finds that these trends are happening across the Western world.
10. See Reeves & Smith (2021) and Reeves et al. (2021).
11. Reeves, R. (2022, October 22). The boys feminism left behind. *Free Press*. www .thefp.com/p/the-boys-feminism-left-behind.
12. I note that in many ways life has gotten better for boys. There has been a big decline in intolerance toward LGBTQ youth, along with a decline in violence of all sorts since the 1980s. There is better mental health treatment and a reduction in the stigma of getting treatment, which used to be particularly strong for boys and men. As Steven Pinker (2011) has shown, life has gotten better in recent centuries in so many ways, for almost everyone, as science advances and rights revolutions proceed. And yet some combination of forces is producing rising numbers of boys who are disengaged from school, work, and family.
13. Reeves (2022, p. xi).
14. See chapter 1 for changes in rates of mental illness among boys since the early 2010s.
15. See Rausch & Haidt (2023, April); Rausch & Haidt (2023, March).
16. See Figures 6.6 on close friends (ch. 6), 6.7 on loneliness (ch. 6), and 7.6 on meaninglessness (ch. 7).

17. Pew Research Center (2019). See graph in the online supplement.
18. U.K. Office for National Statistics (2022).
19. Cai et al. (2023).
20. Reeves & Smith (2020).
21. According to a report published by the Japanese Ministry of Health, Labor, and Welfare, the *hikikomori* are young people who show no interest in personal development or friendship for more than six months, but don't meet the criteria for schizophrenia or other mental disorders. (Ministry of Health, Labor, and Welfare, 2003).
22. Teo & Gaw (2010).
23. Although the research is nuanced and individual differences are necessary to consider, boys (on average) who do not engage in risky play (e.g., rough-and-tumble play or play where children could get lost) are more likely to have difficulties with emotional regulation, social competence, and mental health. See Flanders et al. (2012); Brussoni et al. (2015); See Sandseter, Kleppe, & Sando (2020) for prevalence rates in risky play by gender.
24. See Twenge (2017) for a review.
25. Askari et al. (2022), with data from Monitoring the Future. I thank Melanie Askari for permission to reprint this figure. Zach added the gray shading and line labels. The Y axis converts scale scores to Z scores, which show how high or low a score is in terms of the numbers of standard deviations by which it departs from the zero mark.
26. National Addiction & HIV Data Archive Program. (n.d.-a, n.d.-b). Monitoring the Future.
27. Boys dropped from 49.7% yes in 2010 down to 40.8% in 2019. Girls dropped from 36.4% down to 32.4%. You can find graphs of similar items in the online supplement.
28. Centers for Disease and Control (n.d.). This data set goes back only to 2000.
29. There is a unique exception to the principle that the virtual world pulls boys away from risk in the real world: Social media sometimes incentivizes boys to put themselves and others in danger in order to gain prestige on social media. For instance, viral TikTok challenges often involve dangerous stunts, such as the "Cha Cha Slide" challenge: Participants mimic the song's dance commands while driving, swerving erratically into oncoming traffic. In the "Skull Breaker" challenge, unsuspecting adolescents are tricked into jumping up while their feet are kicked out from under them, leading to severe head injuries and even death. The "Devious Licks" challenge encourages adolescents to live stream themselves vandalizing their school bathrooms. Among the most deadly challenges so far is the "Blackout" challenge, where participants set up a phone to film themselves while they use a rope or other household item to strangle themselves until they fall unconscious. Afterward, they post the video of their blackout and reawakening, for those who do reawaken. In an 18-month span between 2021 and 2022, a *Bloomberg Businessweek* report found that at least 15 children below the age of 12, along with others who were older, had died from this single challenge: Carville, O. (2022, November 30). TikTok's viral challenges keep luring young kids to their deaths. *Bloomberg.* www.bloomberg.com/news/features /2022-11-30/is-tiktok-responsible-if-kids-die-doing-dangerous-viral-challenges. These dangerous challenges are mostly taken on by boys.
30. Orces & Orces (2020).

31. As noted in chapter 3, there is some evidence that depression and anxiety had been slowly increasing among adolescents since the 1940s.

32. Zendle & Cairns (2019); King & Delfabbro (2019); Bedingfield, W. (2022, July 28). It's not just loot boxes: Predatory monetization is everywhere. *Wired*. www.wired.com/story/loot-boxes-predatory-monetization-games.

33. I focus on the dynamics for heterosexual boys because they are the ones whom technology pulls out of sync and far away from the sex they are attracted to. Porn is just as popular with boys who are not heterosexual, although the effects on their sexual development may be different. See Bőthe et al. (2019) for a review of the literature on LGBTQ adolescents and porn, including this line: "LGBTQ adolescents' pornography use does not appear to be related to more negative outcomes compared with heterosexual adolescents; thus, LGBTQ adolescents do not seem more vulnerable to pornographic materials than heterosexual adolescents."

34. Ogas & Gaddam (2011). They note that the number dropped in later years as the internet has increased in website diversity and complexity.

35. Donevan et al. (2022).

36. Donevan et al. (2022).

37. Pizzol et al. (2016).

38. Bőthe et al. (2020).

39. Albright (2008); Szymanski & Stewart-Richardson (2014); Sun et al. (2016). Note that some studies have failed to find this relationship (see Balzarini et al., 2017). Additionally, the relationship between pornography use and relationship quality is complex. Some studies, for example, find that discrepancies in the quantity of porn watched by romantic partners may signal underlying relationship conflict, which then gets exacerbated by pornography use. See Willoughby et al. (2016).

40. Vaillancourt-Morel et al. (2017); Dwulit & Rzymski (2019).

41. Wright et al. (2017).

42. Tolentino, D. (2023, May 12). Snapchat influencer launches an AI-powered "virtual girlfriend" to help "cure loneliness." NBC News. www.nbcnews.com/tech/ai-powered-virtual-girlfriend-caryn-marjorie-snapchat-influencer-rcna84180.

43. See Taylor, J. (2023, July 21). Uncharted territory: Do AI girlfriend apps promote unhealthy expectations for human relationships? *Guardian*. www.theguardian.com/technology/2023/jul/22/ai-girlfriend-chatbot-apps-unhealthy-chatgpt; Murkett, K. (2023, May 12). Welcome to the lucrative world of AI girlfriends. *UnHerd*. unherd.com/thepost/welcome-to-the-lucrative-world-of-ai-girlfriends; Brooks, R. (2023, February 21). I tried the Replika AI companion and can see why users are falling hard. The app raises serious ethical questions. *Conversation*. theconversation.com/i-tried-the-replika-ai-companion-and-can-see-why-users-are-falling-hard-the-app-raises-serious-ethical-questions-200257. Also see India, F. (2023). We can't compete with AI girlfriends. *Girls*. www.freyaindia.co.uk/p/we-cant-compete-with-ai-girlfriends.

44. Fink, E., Segall, L., Farkas, J., Quart, J., Hunt, R., Castle, T., Hottman, A. K., Garst, B., McFall, H., Gomez, G., & BFD Productions. (n.d.). Mostly human: I love you, bot. CNN Money. money.cnn.com/mostly-human/i-love-you-bot/.

45. Su et al. (2020).

46. For evidence that playing violent video games does not cause aggression or violence among users, see Elson & Ferguson (2014); Markey & Ferguson (2017). However, other researchers have found effects between video games use and

aggression, with effect sizes falling around $\beta = .1$. See Bushman & Huesman (2014); Prescott, Sargent & Hull (2016). Also see Anderson et al. (2010).

47. See Alanko (2023) for an extensive review on the social and psychological effects of video game use on adolescents.

48. Kovess-Masfety et al. (2016); Sampalo, Lázaro & Luna (2023).

49. Russoniello et al. (2013).

50. Granic et al. (2014); Greitemeyer & Mügge (2014).

51. Adolescents with certain preexisting mental health conditions are more likely to develop problematic use, for example, than those with preexisting anxiety and/or depression. See Lopes et al. (2022).

52. Pallavicini et al. (2022).

53. The evidence that problematic video game use can ultimately exacerbate loneliness is a topic still debated, and often depends on the role that video games has in a person's life and even the kinds of games that one plays. See Luo et al. (2022).

54. Charlton & Danforth (2007); Lemmens et al. (2009); Brunborg et al. (2013).

55. Young (2009).

56. BBC News. (2022, December 9). Children stopped sleeping and eating to play *Fortnite*—lawsuit. BBC News. www.bbc.com/news/world-us-canada-63911176.

57. See Zastrow (2017); Ferguson et al. (2020).

58. Stevens et al. (2021).

59. Wittek et al. (2016).

60. Brunborg et al. (2013); Fam (2018).

61. The *DSM-5-TR* (American Psychiatric Association, 2022); the diagnosis is still under study. See American Psychiatric Association (2023, January).

62. Chris Ferguson, author of *Moral Combat*, who has been studying the effects of video games on mental health for decades, notes that part of the problem with determining prevalence is that "there is no agreed upon set of symptoms for problematic gaming, nor any single measure of it, so prevalence estimates are all over the place."

63. See Männikkö et al. (2020) for evidence on the effects of problematic gaming on mental health. See also Brailovskaia, Meier-Faust, et al. (2022), who found in an experiment that a two-week period of abstinence from video games reduced stress, anxiety, and other symptoms of internet gaming disorder in a sample of German adults who spent at least three hours per week gaming before the study. Also see Ferguson, Coulson & Barnett (2011), who argue that the evidence between time spent gaming and mental health outcomes is highly varied and may be rooted in underlying mental health problems.

64. Rideout & Robb (2019). Similar results were found in a study of Norwegian adolescents (Brunborg et al., 2013), where the mean time spent gaming per week for girls was 5 hours, and for boys it was 15 hours and 42 minutes. For the addicted gamers, the mean time spent gaming per week was 24 hours.

65. Girls also play video games, but at much lower rates, for less time, with different games, and with less enjoyment, on average, than boys. The 2019 Common Sense report found that 70% of 8-to-18-year-old boys enjoy console gaming "a lot" compared with 23% of girls (Rideout & Robb, 2019). The rates rise for mobile gaming, with 35% of girls reporting high enjoyment compared with 48% of boys. The report also shows that girls spend about 47 minutes per day gaming, with most of that time on their smartphones. Girls, on average, also tend to play

different genres of games from boys, with greater interest in social, puzzle/card, music/dance, educational/edutainment, and simulation games (see Phan et al., 2012; also see Lucas & Sherry, 2004; Lang et al., 2021). And in recent years, there has been an explosive rise in the popularity of female video game streamers, who amass large (mostly male) followings. See Patterson, C. (2023, January 4). Most-watched female Twitch streamers in 2022: Amouranth dominates, VTubers rise up. *Dexerto*. www.dexerto.com/entertainment/most-watched-female-twitch -streamers-in-2022-amouranth-dominates-vtubers-rise-up-2023110.

66. Peracchia & Curcio (2018).
67. Cox (2021).
68. Durkheim (1897/1951, p. 213).

Chapter 8: Spiritual Elevation and Degradation

1. DeSteno (2021).
2. DeSteno's research confirms the 19th-century Danish existentialist philosopher Søren Kierkegaard's (1847/2009) insight that "the function of prayer is not to influence God, but rather to change the nature of the one who prays."
3. See my description of a football game at the University of Virginia in chapter 11 of *The Righteous Mind* (Haidt, 2012).
4. Being "off schedule" with others or not following "temporal norms" in time use surveys predicts being less satisfied with life: Kim (2023).
5. See my discussion of "hive psychology" in chapter 10 of *The Righteous Mind*, which contains many academic citations.
6. See a review of synchrony research, including DeSteno's own research, in the introductory chapter of DeSteno (2021).
7. DeSteno (2021) discusses the importance of sharing food in religious rituals and feasts.
8. The claim that humans evolved to be religious is contested. In *The Righteous Mind*, I explain how religion, morality, and neural circuits for synchrony and self-loss coevolved, drawing on the work of David Sloan Wilson (2002) and many others. But other scholars, such as Richard Dawkins (2006), reject that claim.
9. Eime et al. (2013); Pluhar et al. (2019). Also see Hoffmann et al. (2022). Some portion of this relationship may be reverse correlation—perhaps more sociable kids seek out team sports.
10. Davidson & Lutz (2008).
11. Goyal et al. (2014).
12. Economides et al. (2018).
13. Buchholz (2015); Kenge et al. (2011).
14. Quoted from Maezumi & Cook (2007).
15. Of course, people have been making this charge since the advent of radio and TV. But smartphones and social media demand more attention and create more addictive behavior than portable radios and cassette players (such as the Sony Walkman) ever did.
16. Filipe et al. (2021).
17. Hamilton et al. (2015).
18. See Keltner (2022, p. 37) and Carhart-Harris et al. (2012). For a study showing that awe reduces DMN activity, see van Elk et al. (2019).

19. Keltner (2022, p. 37).

20. See Wang et al. (2023), who found that "individual variations in FOMO are as-sociated with the brain structural architecture of the right precuneus, a core hub within a large-scale functional network resembling the DMN and involved in social and self-referential processes." Maza et al. (2023) conducted a longitu-dinal fMRI study of adolescents going through puberty and found that the brains of heavy social media users changed, over time, compared with light users: Their brains became more sensitive (reactive) to information about im-pending social rewards and punishments.

21. Here I am drawing on Minoura (1992), as well as on research on second-language learning.

22. Berkovitch et al. (2021).

23. Matthew 7:1–2 (NRSV).

24. Matthew 7:3 (NRSV).

25. Seng-ts'an, *Hsin hsin ming*. In Conze (1954).

26. Leviticus 19:18 (NRSV).

27. M. L. King (1957/2012).

28. Dhammapada (Roebuck, 2010).

29. Emerson (1836).

30. Keltner & Haidt (2003). There are many additional perceptions or appraisals that create the many flavors of awe, including threat (as in a thunderstorm or angry deity), beauty, extraordinary or superhuman ability, virtue, and super-natural causality.

31. Tippett, K. (Host). (2023, February 2). Dacher Keltner—the thrilling new science of awe [Audio podcast episode]. *The On Being Project*. onbeing.org/programs/dacher -keltner-the-thrilling-new-science-of-awe.

32. Monroy & Keltner (2023).

33. Wilson (1984).

34. Grassini (2022); Lee et al. (2014).

35. The actual quotation, in translation, is "What else does this craving, and this helplessness, proclaim but that there was once in man a true happiness, of which all that now remains is the empty print and trace? This he tries in vain to fill with everything around him, seeking in things that are not there the help he cannot find in those that are, though none can help, since this infinite abyss can be filled only with an infinite and immutable object; in other words by God him-self." From Pascal (1966, p. 75).

36. Darwin (1871/1998); Wilson (2002).

37. Dhammapada (Roebuck, 2010).

38. Marcus Aurelius (2nd century/2002, p. 59).

Chapter 9: Preparing for Collective Action

1. As Lenore Skenazy did in 2008, earning her the nickname America's Worst Mom.

2. Skenazy (2009).

3. For example, Outsideplay.ca is "a risk reframing tool for caregivers and early childhood educators to manage their fears and develop a plan for change so their children can have more opportunities for risky play." play:groundNYC is

"dedicated to transforming the city through play." They run a fabulous "junk-yard playground" on Governors Island, which my children greatly enjoyed.

4. Sign up at www.afterbabel.com.

Chapter 10: What Governments and Tech Companies Can Do Now

1. Pandey, E. (2017, November 9). Sean Parker: Facebook was designed to exploit human "vulnerability." *Axios*. www.axios.com/2017/12/15/sean-parker-facebook-was-designed-to-exploit-human-vulnerability-1513306782.

2. See Roser et al. (2019) to see the declining trends in child and infant mortality.

3. You can view Harris's presentation at www.minimizedistraction.com.

4. An example of this race to the bottom is TikTok's short-video format, which proved to be highly effective at keeping young people hooked and so was soon copied by Instagram and Facebook Reels, YouTube Shorts, and Snapchat's Spotlight—what Harris calls the TikTokification of social media. I thank Jamie Neikrie for this example.

5. Harris, T. Retrieved from www.commerce.senate.gov/services/files/96E3A739-DC8D-45F1-87D7-EC70A368371D.

6. See the age verification section of Social Media Reform: A Collaborative Review, available at www.anxiousgeneration.com/reviews.

7. Heath, A. (2021, October 15). Facebook's lost generation. *Verge*. www.theverge.com/22743744/facebook-teen-usage-decline-frances-haugen-leaks.

8. Wells, G. & Horwitz, J. (2021, September 28). Facebook's effort to attract pre-teens goes beyond Instagram kids, documents show. *Wall Street Journal*. www.wsj.com/articles/facebook-instagram-kids-tweens-attract-11632849667.

9. Meta. (2023, June 29). Instagram Reels Chaining AI system. www.transparency.fb.com/features/explaining-ranking/ig-reels-chaining/?referrer=1.

10. Hanson, L. (2021, June 11). Asking for a friend: What if the TikTok algorithm knows me better than I know myself? *GQ Australia*. www.gq.com.au/success/opinions/asking-for-a-friend-what-if-the-tiktok-algorithm-knows-me-better-than-i-know-myself/news-story/4eea6d6f23f9ead544c2f773c9a13921; Barry, R., Wells, G., West, J., Stern, J., & French, J. (2021, September 8). How TikTok serves up sex and drug videos to minors. *Wall Street Journal*. www.wsj.com/articles/tiktok-algorithm-sex-drugs-minors-11631052944.

11. The Data Team (2018, May 18). How heavy use of social media is linked to mental illness. *The Economist*. www.economist.com/graphic-detail/2018/05/18/how-heavy-use-of-social-media-is-linked-to-mental-illness.

12. The law is unlikely to take effect for many years, if ever. The platforms are blocking implementation of design codes in multiple states with lawsuits alleging that most of the provisions of the AADC violate the first amendment of the United States constitution. The platforms are essentially arguing that they cannot be regulated because any regulation would have some effect on the speech carried out on the platform.

13. Zach and I are collaborating with the Center for Humane Technology to collect and analyze the many approaches being proposed or implemented by governments and legislatures in the United States and other countries. You can find the link at www.anxiousgeneration.com/reviews. Also see Rausch & Haidt (2023, November).

14. Newton, C. (2023, August 4). How the kids online safety act puts us all at risk.

The Verge. www.theverge.com/2023/8/4/23819578/kosa-kids-online-safety-act -privacy-danger. For another example, see: The Free Press (2022, December 15). Twitter's secret blacklists. The Free Press. www.thefp.com/p/twitters-secret -blacklists.

15. For a fuller discussion of the limits of content moderation, see Iyer, R. (2022, October 7). Content moderation is a dead end. *Designing Tomorrow*, Substack. psychoftech.substack.com/p/content-moderation-is-a-dead-end.

16. For a fuller discussion of platform design, including several examples, see Howell, J. P., Jurecic, Q., Rozenshtein, A. Z., & Iyer, R. (2023, March 27). Ravi Iyer on how to improve technology through design. *The Lawfare Podcast*. www.lawfaremedia .org/article/lawfare-podcast-ravi-iyer-how-improve-technology-through-design.

17. Evans, A., & Sharma, A. (2021, August 12). Furthering our safety and privacy commitments for teens on TikTok. TikTok. newsroom.tiktok.com/en-us/furthering -our-safety-and-privacy-commitments-for-teens-on-tiktok-us.

18. Instagram. (2021, July 27). Giving young people a safer, more private experience. Instagram. about.instagram.com/blog/announcements/giving-young-people-a -safer-more-private-experience.

19. They are also language neutral, whereas mandates to perform more content moderation are unlikely to be well implemented in nearly all of the hundreds of languages that Facebook supports. Frances Haugen has been outspoken about the value of design changes that can be implemented across all languages easily.

20. Jargon, J. (2019, June 18). How 13 became the internet's age of adulthood. *Wall Street Journal*. www.wsj.com/articles/how-13-became-the-internets-age-of-adulthood -11560850201.

21. See endnote 20.

22. Orben et al. (2022).

23. Even if an age verification company got hacked, as long as they stored their data thoughtfully, there would be nothing linking their customers to any particular site that had asked about them.

24. The Age Verification Providers Association, avpassociation.com.

25. See here for the ways that Meta has begun to offer more age verification options for users: Meta. (2022, June 23). Introducing new ways to verify age on Instagram. *Meta*. www.about.fb.com/news/2022/06/new-ways-to-verify-age-on-instagram.

26. The next generation of the internet can and should be built so that people control their own data and can decide how it is used. See ProjectLiberty.io for one such vision.

27. Parents can use monitoring and filtering programs, in conjunction with their home router, to accomplish such blocking. I will say more about such programs on my Substack. But these are somewhat complicated steps that parents must take, which means that they will only be used by a small subset of parents. I am proposing defaults that would apply automatically, unless the parent specifically changes the default setting.

28. Skenazy, L. (2014, July 14). Mom jailed because she let her 9-year-old daughter play in the park unsupervised. *Reason*. www.reason.com/2014/07/14/mom-jailed -because-she-let-her-9-year-ol.

29. Skenazy, L. (2022, December 8). CPS: Mom can't let her 3 kids—ages 6, 8, and 9—play outside by themselves. *Reason*. www.reason.com/2022/12/08/emily-fields -pearsiburg-virginia-cps-kids-outside-neglect.

30. St. George, D. (2015, June 22). "Free range" parents cleared in second neglect case after kids walked alone. *Washington Post*. www.washingtonpost.com/local /education/free-range-parents-cleared-in-second-neglect-case-after-children -walked-alone/2015/06/22/82283c24-188c-11e5-bd7f-4611a60dd8e5_story.html.

31. Flynn et al. (2023).

32. Mom issued misdemeanor for leaving 11-year-old in car. (2014, July 9). NBC Connecticut. www.nbcconnecticut.com/news/local/mom-issued-misdemeanor-for -leaving-11-year-old-in-car/52115.

33. For those interested in helping their state (or even city or town) pass a Reasonable Childhood Independence bill, Let Grow has a free legislative action "toolkit" on its site: www.letgrow.org/legislative-toolkit.

34. See Free Play and Mental Health: A Collaborative Review, at www.jonathanhaidt .com/reviews.

35. The U.S. Centers for Disease Control recommends that all grades get recess, even high school. See Centers for Disease Control (n.d.). Recess. CDC Healthy Schools. www.cdc.gov/healthyschools/physicalactivity/recess.htm.

36. Young et al. (2023).

37. Sanderson, N. (2019, May 30). What are school streets? 8 80 Cities. www.880cities .org/what-are-school-streets.

38. Another way cities can be more child-friendly is by making public transit affordable and welcoming to kids. Tim Gill, author of "Urban Playground: How Child-Friendly Planning and Design Can Save Cities," notes that in London children aged 5–10 can travel for free, without an adult, on the tube and buses.

39. See review of research in Reeves (2022, Chapter 10).

40. As an example: In the United States, the National Apprenticeship Act would invest $3.5 billion over five years to create nearly a million new apprenticeships for young people.

41. Bowen et al. (2016); Gillis et al. (2016); Bettmann et al. (2016); Wilson & Lipsey (2000); Beck & Wong (2022); Davis-berman & Berman (1989); Gabrielsen et al. (2019); Stewart (1978).

42. The DCF Wilderness School, portal.ct.gov/DCF/Wilderness-School/Home. Other states run similar programs. See the Montana Wilderness School, www .montanawildernessschool.org.

Chapter 11: What Schools Can Do Now

1. St. George, D. (2023, April 28). One school's solution to the mental health crisis: Try everything. *Washington Post*. www.washingtonpost.com/education/2023 /04/28/school-mental-health-crisis-ohio.

2. Brundin, J. (2019, November 5). This Colorado middle school banned phones 7 years ago. They say students are happier, less stressed, and more focused. Colorado Public Radio. www.cpr.org/2019/11/05/this-colorado-middle-school-banned -phones-seven-years-ago-they-say-students-are-happier-less-stressed-and-more -focused.

3. The phone policy works like this: "There's a warning the first time a phone is out of a student's backpack. On the second infraction, the phone is confiscated and parents have to pick it up. The third time, a student must hand the phone into

the office at the beginning of the school day and pick it up at the end, for a set period of time."

4. Walker, T. (2023, February 3). Cellphone bans in school are back. How far will they go? *NEA Today.* www.nea.org/advocating-for-change/new-from-nea/cellphone -bans-school-are-back-how-far-will-they.go.

5. In 2023, the American Federation of Teachers issued a report calling out social media platforms for "undermining classroom learning, increasing costs for school systems and being a 'root cause' of the nationwide youth mental health crisis." See American Federation of Teachers. (2023, July 20). New report calls out social media platforms for undermining schools, increasing costs, driving youth mental health crisis. www.aft.org/press-release/new-report-calls-out -social-media-platforms-undermining-schools-increasing-costs.

6. See the quotation from Ken Trump in this essay: Walker, T. (2023, February 3). Cellphone bans in school are back. How far will they go? *NEA Today.* www.nea .org/advocating-for-change/new-from-nea/cellphone-bans-school-are-back -how-far-will-they-go. It is also relevant that the middle school in Newtown, Connecticut, where a horrific school shooting took place in an elementary school in 2012, decided in 2022 to require students to keep their phones in their lockers all day long. A parent pointed me to their parent handbook: Newtown Public School District. (n.d.). *Newtown middle school, 2022–2023 student/parent handbook.* nms.newtown.k12.ct.us/_theme/files/2022-2023/2022-2023%20Student _Parent%20Handbook_docx.pdf.

7. See UNESCO (2023). *Technology in education: A tool on whose terms?* www.unesco .org/gem-report/en/technology. See a summary of the phone recommendations here: Butler, P., & Farah, H. (2023, July 25). "Put learners first": Unesco calls for global ban on smartphones in schools. *Guardian.* www.theguardian.com/world /2023/jul/26/put-learners-first-unesco-calls-for-global-ban-on-smartphones-in -schools.

8. Zach Rausch and I have been collecting the evidence related to phone-free schools in a collaborative review document, available at www.jonathanhaidt .com/reviews.

9. Richtel, M. (2011, October 22). A Silicon Valley school that doesn't compute. *New York Times.* www.nytimes.com/2011/10/23/technology/at-waldorf-school-in -silicon-valley-technology-can-wait.html; Bowles, N. (2018, October 26). The digital gap between rich and poor kids is not what we expected. *New York Times.* www .nytimes.com/2018/10/26/style/digital-divide-screens-schools.html.

10. See graphs in the online supplement, or here: National Center for Education Statistics (n.d.). The drop in scores from the 2020 academic year (measured before the COVID shutdowns) to the 2022 academic year was 9 points in math and 4 points in reading. The drop from 2012 to 2020 was 5 points in math and 3 points in reading.

11. Twenge, Wang, et al. (2022). See also Nagata, Singh et al. (2022).

12. I know of no school district that has tested the hypothesis experimentally by randomly assigning middle schools to go phone-free while others make no change. This is the most important study I can think of to address the mental health crisis. I say more about what such a study would look like in Social Media and Mental Health: A Collaborative Review, available in the online supplement.

13. See "Khanmingo," Khan Academy's personal AI assistant: Khan Academy. (n.d.). *World-class AI for education*. www.khanacademy.org/khan-labs.

14. Stinehart, K. (2021, November 23). Why unstructured free play is a key remedy to bullying. *eSchoolNews*. www.eschoolnews.com/sel/2021/11/23/why-unstructured -free-play-is-a-key-remedy-to-bullying.

15. For a longer list of suggestions and updates to this list, see the online supplement.

16. All of Let Grow's materials are free. See www.letgrow.org/program/the-let-grow -project.

17. Soave, R. (2014, November 20). Schools to parents: Pick up your kids from the bus or we'll sic child services on you. *Reason*. reason.com/2014/11/20/child-services -will-visit-parents-who-le.

18. Skenazy, L. (2016, November 7). Local library will call the cops if parents leave their kids alone for 5 minutes. *Reason*. reason.com/2016/11/07/local-library-will -call-the-cops-if-pare.

19. Centers for Disease Control and Prevention (2015, p. 134).

20. See Martinko, K. (2018, October 11). Children spend less time outside than prison inmates. *Treehugger*. www.treehugger.com/children-spend-less-time-outside -prison-inmates-4857353. See also for research behind this claim Edelman, R. (2016, April 4). Dirt is good: The campaign for play. *Edelman*. www.edelman.co.uk /insights/dirt-good-campaign-play.

21. The report was issued by the U.S. National Commission on Excellence in Education. See Gray et al. (2023).

22. The 2001 federal No Child Left Behind Act was a major spur to the focus on test scores. The Common Core State Standards were developed in 2009 and released in 2010. Adoption was rapid, with 45 states and the District of Columbia adopting the standards. However, five of these states later repealed or replaced those standards. See Common Core implementation by state. *Wikipedia*. en.wikipedia .org/wiki/Common_Core_implementation_by_state.

23. Atlanta public schools cheating scandal. Wikipedia. en.wikipedia.org/wiki /Atlanta_Public_Schools_cheating_scandal.

24. Murray & Ramstetter (2013). Also see Singh et al. (2012) for research on the link between physical activity and school performance.

25. Haapala et al. (2016).

26. Centers for Disease Control (2017, January). Strategies for recess in schools. U.S. Department of Health and Human Services. www.cdc.gov/healthyschools /physicalactivity/pdf/2019_04_25_SchoolRecess_strategies_508tagged.pdf.

27. Brooklyn Bridge Parents (2017, May 7). A look inside the junk yard playground on Governors Island. brooklynbridgeparents.com/a-look-inside-the-junk-yard -playground-on-governors-island.

28. Keeler (2020).

29. Photo by Jonathan Haidt.

30. I thank Adam Bienenstock for this photograph. Bienenstock built the playground using a design from the Danish playground architect Helle Nebelong.

31. Fyfe-Johnson et al. (2021).

32. Vella-Brodrick & Gilowska (2022).

33. Lahey, J. (2014, January 28). Recess without rules. *Atlantic*. www.theatlantic.com /education/archive/2014/01/recess-without-rules/283382; see also Saul, H. (2014,

January 28). New Zealand school bans playground rules and sees less bullying and vandalism. *Independent*. www.independent.co.uk/news/world/australasia /new-zealand-school-bans-playground-rules-and-sees-less-bullying-and -vandalism-9091186.html.

34. See endnote 33.

35. Brussoni et al. (2017).

36. Healthy play is not entirely painless. Roughhousing, name-calling, scrapes, and bruises are all parts of natural play, and they are necessary for the antifragility effects of play to take hold. Removing them from recess to keep kids "safe" is like removing all the nutrients from wheat and feeding kids only white bread. I'm not saying that we should ever accept bullying. Bullying by most definitions requires a repeated pattern, over more than one day, of one child intending to harm another. There is an essential role for adults in setting up policies to reduce bullying, and in responding to it when it happens. But the vast majority of conflicts and cases of teasing and name-calling are not bullying, and adults should not rush in to stop them.

37. Dee (2006); Mullola et al. (2012).

38. Partelow (2019, p. 3).

39. See Reeves (2022, September); Casey and Nzau (2019); Torre (2018).

40. These two paragraphs were taken from an essay I wrote in *The Atlantic*, where I expanded on the case for phone-free schools: Haidt, J. (2023, June 6). Get phones out of schools now. *Atlantic*. www.theatlantic.com/ideas/archive/2023/06/ban -smartphones-phone-free-schools-social-media/674304.

41. I recognize that schools may face increased liability risks and insurance costs. I hope that governments can pass liability reforms that would free schools to focus on education rather than lawsuits. See Howard (2014) for a discussion of how this can be done.

42. This would rectify one of the biggest shortcomings in the scientific literature: the focus on individual-level effects because there is almost no research on entire schools that went phone-free or that greatly expanded free play and autonomy. Let's measure the emergent group-level effects of these policies.

43. If the number of schools is very large, then random assignment would work well. If there are only 16 middle schools, let's say, and if they vary by race or social class, then it would be wise to separate similar schools to be sure that each of the four groups is as comparable to the others as possible. Once the groups are created, the assignment of each group to an experimental condition could be done using a random method such as drawing numbers from a hat.

44. Before the experiment begins, a set of agreed-upon measures are collected, or created if they don't exist, to measure the key variables that the school cares about, such as academic achievement, mental health referrals, student reports of their mental health and their engagement with school, measures of bullying and behavioral problems, and teacher reports about classroom culture including students' ability to stay on task and engage with the lesson. The measures would be collected every month, if possible, or at least three times during the school year.

45. In elementary schools, where phones may not yet be much of a problem, a district might want to try simpler versions of the experiment with just two conditions: Play Club versus no Play Club, or Let Grow Project versus no Let Grow

Project. Experiments like these should be tried in different regions and countries to see how well they work in varied conditions.

Chapter 12: What Parents Can Do Now

1. Gopnik (2016, p. 18).
2. Lenore Skenazy wrote sections of this chapter with me, drawing on her experience as president of LetGrow.org, which she and I cofounded in 2017, along with Peter Gray and Daniel Shuchman. For a more comprehensive list of suggestions and post-publication updates, see the online supplement for this chapter, and also www.letgrow.org.
3. Scarr (1992).
4. See the online appendix for the rising rates of U.S. 12th graders who believe that their "life is not very useful."
5. For summaries and links to such recommendations, see my collaborative review document: The Impact of Screens on Infants, Toddlers, and Preschoolers, www.anxiousgeneration.com/reviews.
6. Myers et al. (2017); Kirkorian & Choi (2017); Roseberry et al. (2014).
7. At least the advice is consistent from medical authorities in the United States (Council on Communications and Media, 2016); Canada (Ponti et al., 2017); and Australia (Joshi & Hinkley, 2021). The U.K. is somewhat more lax (Viner et al., 2019).
8. These bullet points are directly quoted from the American Academy of Child & Adolescent Psychiatry (2020).
9. I would just note that watching shows and movies for part of long car trips and plane trips seems fine for children over two or three.
10. Harris (1989).
11. Let Grow (n.d.). Kid license, www.letgrow.org/printable/letgrowlicense.
12. Safe Routes to School, www.saferoutesinfo.org.
13. For a list of summer camps in the United States and Canada that show evidence of supporting free-range childhoods, see Skenazy, L. (2023, August 14). Phone-free camps. Let Grow. www.letgrow.org/resource/phone-free-camps.
14. Be sure the camp truly locks up phones for the duration; many camps say that they ban phones, but like many schools they really just mean "Don't let an adult see you taking your phone out of your pocket."
15. See this profile of playborhoods: Thernstrom, M. (2016, October 16). The anti-helicopter parent's plea: Let kids play! *New York Times.* www.nytimes.com/2016/10/23/magazine/the-anti-helicopter-parents-plea-let-kids-play.html. Also see Lanza's book and website at www.playborhood.com.
16. Some parents are concerned that they may be liable if someone else's child gets hurt. The fear of lawsuits can be paralyzing. But Lanza says he decided against waivers or getting extra insurance. Instead, he removed any obvious dangers from his yard, added some play equipment—he recommends things like swings, a playhouse, some area to make art—and trusted his neighbors not to sue him. It worked. Kids there are getting a taste of a play-based childhood. A no-cost and lower liability risk option is to have parents rotate responsibility for being present at a local park or playground each afternoon, so that families know that there will be an adult on hand who will only get involved in case of emergencies.
17. Lin, H. (2023). Your First Device. *The New Yorker* © Condé Nast.

18. Children in this age range are noticing and copying adult behavior; therefore, it is important to model healthy technology use. You don't have to be perfect, but do try to show your children how to set healthy boundaries so that they see you making an effort to keep screens in their proper place, and to be fully present when it is time to be fully present. For practical guidance on how to model technology use for your children, see Nelson (2023, September 28). How Parents Can Model Appropriate Digital Behavior for Kids. www.brightcanary.io/parents-digital-role-model.

19. Rideout (2021).

20. Nesi (2023).

21. Specifically, Knorr, C. (2021, March 9). Parents' ultimate guide to parental controls. Common Sense Media. www.commonsensemedia.org/articles/parents-ultimate-guide-to-parental-controls.

22. See Sax, L. (2022, September 7). Is your son addicted to video games? Institute for Family Studies. ifstudies.org/blog/is-your-son-addicted-to-video-games. Sax also suggests that parents use Common Sense Media to understand the games that their children are playing. Just type in the name of the game, and the site provides a summary of the game and the age range for which the game is suitable.

23. Melanie Hempe of ScreenStrong urges parents to not allow *any* devices in bedrooms. She tells me that "the majority of dark screen activities happen behind a closed bedroom door."

24. See, for example, www.healthygamer.gg, www.gamequitters.com, and www.screenstrong.org.

25. See the Screen Time Action Network from FairPlay. www.screentimenetwork.org.

26. Kremer, W. (2014, March 23). What medieval Europe did with its teenagers. BBC. www.bbc.com/news/magazine-26289459.

27. American Exchange Project, americanexchangeproject.org/about-us.

28. American Field Service, www.afsusa.org/study-abroad.

29. See the online supplement for more links, and for programs not based in the United States.

30. CISV International, cisv.org/about-us/our-story.

31. There are a number of additional outdoor expedition programs for teens throughout the United States including programs through the YMCA (see ycamp.org/wilderness-trips and www.ymcanorth.org/camps/camp_menogyn/summer_camp), Wilderness Adventures (www.wildernessadventures.com), Montana Wilderness School (www.montanawildernessschool.org), NOLS (nols.edu/en), and Outward Bound (www.outwardbound.org).

32. See the DCF Wilderness school at portal.ct.gov/DCF/Wilderness-School/Home.

33. I list more such sites in the online supplement.

34. See Center for Humane Technology (n.d.). Youth toolkit. www.humanetech.com/youth. See also Screensense at www.screensense.org, and Screen Time Action Network from Fairplay. www.screentimenetwork.org.

Conclusion: Bring Childhood Back to Earth

1. See Alternative Hypotheses to the Adolescent Mental Illness Crisis: A Collaborative Review, available at www.anxiousgeneration.com/reviews.

2. Middle schools should see substantial improvements within two years, given

that these four reforms would all make daily life more playful and sociable, and less phone-based. In high schools, it will be hard for parents to get students off social media when they are already on. High schools are likely to see some immediate benefits from going phone free. But the biggest improvements might not be seen until several years of new students have entered the high school from middle schools and families that had delayed the phone-based childhood until high school.

3. Latane & Darley (1968). There was a third condition, in which the real subject was in the waiting room with two other students who were working for the experimenters. Their job was to just sit there and keep filling out their questionnaires. In that condition, only 10% of the students got up to report the smoke. This is the finding usually presented in discussions of this study, but I think the condition with three real subjects is the most important one.

4. Titanium dioxide causes a variety of harms; this experiment would never be done today. (The researchers likely did not know of the harmful effects at the time.)

5. See www.letgrow.org, www.outsideplay.ca, www.fairplayforkids.org, and others that I list in the online supplement at www.anxiousgeneration.com/supplement. I admire the approach taken by www.waituntil8th.org, but I think they should change their name to WaitUntil9th.org.

6. See www.humanetech.com, www.commonsense.org, www.screenstrong.org, www.screensense.org, and others that I list in the online supplement.

7. See www.designitforus.org, and other organizations I list in the online supplement.

8. Keaggy, D. T. (2023, February 14). Lembke testifies before Senate committee on online safety. *The Source—Washington University in St. Louis*. Retrieved from www .source.wustl.edu/2023/02/lembke-testifies-before-senate-committee-on-online -safety.

REFERENCES

Ahmed, G. K., Abdalla, A. I., Mohamed, A. W., Mohamed, L. K., & Shamaa, H. A. (2022). Relationship between time spent playing internet gaming apps and behavioral problems, sleep problems, alexithymia, and emotion dysregulations in children: A multicentre study. *Child and Adolescent Psychiatry and Mental Health*, *16*, Article 67. doi.org/10.1186/s13034-022-00502-w

Aichner, T., Grünfelder, M., Maurer, O., & Jegeni, D. (2021). Twenty-five years of social media: A review of social media applications and definitions from 1994 to 2019. *Cyberpsychology, Behavior, and Social Networking*, *24*(4), 215–222. doi.org/10.1089/cyber.2020.0134

Alanko, D. (2023). The health effects of video games in children and adolescents. *Pediatrics In Review*, *44*(1), 23–32. doi.org/10.1542/pir.2022-005666

Aitken, M., Steensma, T. D., Blanchard, R., VanderLaan, D. P., Wood, H., Fuentes, A., & Zucker, K. J. (2015). Evidence for an altered sex ratio in clinic-referred adolescents with gender dysphoria. *The Journal of Sexual Medicine*, *12*(3), 756–763. doi.org/10.1111/jsm.12817

Alavi, S. S., Ferdosi, M., Jannatifard, F., Eslami, M., Alaghemandan, H., & Setare, M. (2012). Behavioral addiction versus substance addiction: Correspondence of psychiatric and psychological views. *International Journal of Preventive Medicine*, *3*(4), 290–294.

Albright, J. M. (2008). Sex in America online: An exploration of sex, marital status, and sexual identity in internet sex seeking and its impacts. *Journal of Sex Research*, *45*(2), 175–186. doi.org/10.1080/00224490801987481

Alimoradi, Z., Lin, C.-Y., Broström, A., Bülow, P. H., Bajalan, Z., Griffiths, M. D., Ohayon,

M. M., & Pakpour, A. H. (2019). Internet addiction and sleep problems: A systematic review and meta-analysis. *Sleep Medicine Reviews, 47,* 51–61. doi.org/10.1016/j.smrv.2019.06.004

Alonzo, R., Hussain, J., Stranges, S., & Anderson, K. K. (2021). Interplay between social media use, sleep quality, and mental health in youth: A systematic review. *Sleep Medicine Reviews, 56,* 101414. doi.org/10.1016/j.smrv.2020.101414

American Academy of Child & Adolescent Psychiatry. (2020, February). *Screen time and children.* www.aacap.org/AACAP/Families_and_Youth/Facts_for_Families/FFF-Guide/Children-And-Watching-TV-054.aspx

American College Health Association (n.d.). *National College Health Assessment.* www.acha.org/NCHA/About_ACHA_NCHA/Survey/NCHA/About/Survey.aspx?hkey=7e9f6752-2b47-4671-8ce7-ba7a529c9934

American College of Pediatricians. (2020, May). *Media use and screen time—its impact on children, adolescents, and families.* acpeds.org/position-statements/media-use-and-screen-time-its-impact-on-children-adolescents-and-families

American Psychiatric Association. (2022, March). *Diagnostic and statistical manual of mental disorders* (5th ed., text rev.). doi.org/10.1176/appi.books.9780890425787

American Psychiatric Association. (2023, January). Internet gaming. www.psychiatry.org/patients-families/internet-gaming

Ames, L. B., & Ilg, F. L. (1979). *Your six-year-old: Defiant but loving.* Delacorte Press.

Anderson, C. A., Shibuya, A., Ihori, N., Swing, E. L., Bushman, B. J., Sakamoto, A., Rothstein, H. R., & Saleem, M. (2010). Violent video game effects on aggression, empathy, and prosocial behavior in Eastern and Western countries: A meta-analytic review. *Psychological Bulletin, 136*(2), 151–173. doi.org/10.1037/a0018251

Archer, J. (2004). Sex differences in aggression in real-world settings: A meta-analytic review. *Review of General Psychology, 8*(4), 291–322. doi.org/10.1037/1089-2680.8.4.291

Arenas-Arroyo, E., Fernández-Kranz, D., & Nollenberger, N. (2022). High speed internet and the widening gender gap in adolescent mental health: Evidence from hospital records. *IZA Discussion Papers,* No. 15728. www.iza.org/publications/dp/15728/high-speed-internet-and-the-widening-gender-gap-in-adolescent-mental-health-evidence-from-hospital-records

Askari, M. S., Rutherford, C., Mauro, P. M., Kreski, N. T., & Keyes, K. M. (2022). Structure and trends of externalizing and internalizing psychiatric symptoms and gender differences among adolescents in the US from 1991 to 2018. *Social Psychiatry and Psychiatric Epidemiology, 57*(4), 737–748. doi.org/10.1007/s00127-021-02189-4

Assari, S. (2020). American children's screen time: Diminished returns of household income in Black families. *Information, 11*(11), 538. doi.org/10.3390/info11110538

Atske, S., & Perrin, A. (2021, July 16). Home broadband adoption, computer ownership vary by race, ethnicity in the U.S. Pew Research Center. www.pewresearch.org/short-reads/2021/07/16/home-broadband-adoption-computer-ownership-vary-by-race-ethnicity-in-the-u-s/

Australian Institute of Health and Welfare. (2022). *Australia's health snapshots 2022: Mental health of young Australians.* www.aihw.gov.au/getmedia/ba6da461-a046-44ac-9a7f-29d08a2bea9f/aihw-aus-240_Chapter_8.pdf.aspx

Auxier, M., Anderson, M., Perrin, A., & Turner, E. (2020, July 28). Parenting children in the age of screens. Pew Research Center. www.pewresearch.org/internet/2020/07/28/parenting-children-in-the-age-of-screens/

Balzarini, R. N., Dobson, K., Chin, K., & Campbell, L. (2017). Does exposure to erotica reduce attraction and love for romantic partners in men? Independent replications of Kenrick, Gutierres, and Goldberg (1989) study 2. *Journal of Experimental Social Psychology, 70*, 191–197. doi.org/10.1016/j.jesp.2016.11.003

Barrick, E. M., Barasch, A., & Tamir, D. I. (2022). The unexpected social consequences of diverting attention to our phones. *Journal of Experimental Social Psychology, 101*, 104344. doi.org/10.1016/j.jesp.2022.104344

Bauer, M., Blattman, C., Chytilová, J., Henrich, J., Miguel, E., & Mitts, T. (2016). Can war foster cooperation? *Journal of Economic Perspectives, 30*(3), 249–274. doi.org/10.1257/jep.30.3.249

Baumgartner, S. E., van der Schuur, W. A., Lemmens, J. S., & te Poel, F. (2018). The relationship between media multitasking and attention problems in adolescents: Results of two longitudinal studies. *Human Communication Research, 44*(1), 3–30. doi.org/10.1093/hcre.12111

Beck, N., & Wong, J. S. (2022). A meta-analysis of the effects of wilderness therapy on delinquent behaviors among youth. *Criminal Justice and Behavior, 49*(5), 700–729. doi.org/10.1177/00938548221078002

Berger, M. N., Taba, M., Marino, J. L., Lim, M. S. C., Cooper, S. C., Lewis, L., Albury, K., Chung, K. S. K., Bateson, D., & Skinner, S. R. (2021). Social media's role in support networks among LGBTQ adolescents: A qualitative study. *Sexual Health, 18*(5), 421–431. doi.org/10.1071/SH21110

Berger, M. N., Taba, M., Marino, J. L., Lim, M. S. C., & Skinner, S. R. (2022). Social media use and health and well-being of lesbian, gay, bisexual, transgender, and queer youth: Systematic review. *Journal of Medical Internet Research, 24*(9), Article e38449. doi.org/10.2196/38449

Bettmann, J. E., Gillis, H. L., Speelman, E. A., Parry, K. J., & Case, J. M. (2016). A meta-analysis of wilderness therapy outcomes for private pay clients. *Journal of Child and Family Studies, 25*(9), 2659–2673. doi.org/10.1007/s10826-016-0439-0

Bijttebier, P., Beck, I. M., Claes, L., & Vandereycken, W. (2009). Gray's reinforcement sensitivity theory as a framework for research on personality–psychopathology associations. *Clinical Psychology Review, 29*(5), 421–430. doi.org/10.1016/j.cpr.2009.04.002

Black, J. E., Jones, T. A., Nelson, C. A., & Greenough, W. T. (1998). Neuronal plasticity and the developing brain. In *Handbook of Child and Adolescent Psychiatry* (Vol. 6, pp. 31–53).

Block, J. (2023). Gender dysphoria in young people is rising—and so is professional disagreement. *BMJ, 380*, 382. doi.org/10.1136/bmj.p382

Boer, M., Stevens, G., Finkenauer, C., & van den Eijnden, R. (2019). Attention deficit hyperactivity disorder-symptoms, social media use intensity, and social media use problems in adolescents: Investigating directionality. *Child Development, 91*(4), e853–e865. doi.org/10.1111/cdev.13334

Borca, G., Bina, M., Keller, P. S., Gilbert, L. R., & Begotti, T. (2015). Internet use and developmental tasks: Adolescents' point of view. *Computers in Human Behavior, 52*, 49–58. doi.org/10.1016/j.chb.2015.05.029

Boss, L. P. (1997). Epidemic hysteria: A review of the published literature. *Epidemiologic Reviews, 19*(2), 233–243. doi.org/10.1093/oxfordjournals.epirev.a017955

Botella, C., Fernández-Álvarez, J., Guillén, V., García-Palacios, A., & Baños, R. (2017).

Recent progress in virtual reality exposure therapy for phobias: A systematic review. *Current Psychiatry Reports, 19*(7), Article 42. doi.org/10.1007/s11920-017-0788-4

Bőthe, B., Vaillancourt-Morel, M.-P., Bergeron, S., & Demetrovics, Z. (2019). Problematic and non-problematic pornography use among LGBTQ adolescents: A systematic literature review. *Current Addiction Reports, 6*, 478–494. doi.org/10.1007/s40429-019-00289-5

Bőthe, B., Vaillancourt-Morel, M.-P., Girouard, A., Štulhofer, A., Dion, J., & Bergeron, S. (2020). A large-scale comparison of Canadian sexual/gender minority and heterosexual, cisgender adolescents' pornography use characteristics. *Journal of Sexual Medicine, 17*(6). doi.org/10.1016/j.jsxm.2020.02.009

Bowen, D. J., Neill, J. T., & Crisp, S. J. R. (2016). Wilderness adventure therapy effects on the mental health of youth participants. *Evaluation and Program Planning, 58*, 49–59. doi.org/10.1016/j.evalprogplan.2016.05.005

Boyd, R., & Richerson, P. J. (1985). *Culture and the evolutionary process.* University of Chicago Press.

Brady, W. J., Wills, J. A., Jost, J. T., Tucker, J. A., & Van Bavel, J. J. (2017). Emotion shapes the diffusion of moralized content in social networks. *Proceedings of the National Academy of Sciences of the United States of America, 114*(28), 7313–7318. doi.org/10.1073/pnas.1618923114

Braghieri, L., Levy, R., & Makarin, A. (2022). Social media and mental health. *American Economic Review, 112*(11), 3660–3693. doi.org/10.1257/aer.20211218

Brailovskaia, J., Krasavtseva, Y., Kochetkov, Y., Tour, P., & Margraf, J. (2022). Social media use, mental health, and suicide-related outcomes in Russian women: A cross-sectional comparison between two age groups. *Women's Health, 18.* doi.org/10.1177/17455057221141292

Brailovskaia, J., Meier-Faust, J., Schillack, H., & Margraf, J. (2022). A two-week gaming abstinence reduces internet gaming disorder and improves mental health: An experimental longitudinal intervention study. *Computers in Human Behavior, 134.* doi.org/10.1016/j.chb.2022.107334

Brand, B. L., Sar, V., Stavropoulos, P., Krüger, C., Korzekwa, M., Martínez-Taboas, A., & Middleton, W. (2016). Separating fact from fiction: An empirical examination of six myths about dissociative identity disorder. *Harvard Review of Psychiatry, 24*(4), 257–270. doi.org/10.1097/hrp.0000000000000100

Brown, D. (1991). *Human universals.* McGraw-Hill.

Brunborg, G. S., Mentzoni, R. A., Melkevik, O. R., Torsheim, T., Samdal, O., Hetland, J., Andreassen, C. S., & Palleson, S. (2013). Gaming addiction, gaming engagement, and psychological health complaints among Norwegian adolescents. *Media Psychology, 16*(1), 115–128. doi.org/10.1080/15213269.2012.756374

Brussoni, M., Gibbons, R., Gray, C., Ishikawa, T., Sandseter, E. B. H., Bienenstock, A., Chabot, G., Fuselli, P., Herrington, S., Janssen, I., Pickett, W., Power, M., Stanger, N., Sampson, M., & Tremblay, M. S. (2015). What is the relationship between risky outdoor play and health in children? A Systematic Review. *International Journal of Environmental Research and Public Health, 12*(6), 6423–6454. doi.org/10.3390/ijerph120606423

Brussoni, M., Ishikawa, T., Brunelle, S., & Herrington, S. (2017). Landscapes for play: Effects of an intervention to promote nature-based risky play in early childhood centres. *Journal of Environmental Psychology, 54*, 139–150. doi.org/10.1016/j.jenvp.2017.11.001

Brussoni, M., Olsen, L. L., Pike, I., & Sleet, D. A. (2012). Risky play and children's safety: Balancing priorities for optimal child development. *International Journal of Environmental Research and Public Health*, *9*(9), 3134–3148. doi.org/10.3390/ijerph9093134

Buchholz, L. (2015). Exploring the promise of mindfulness as medicine. *JAMA*, *314*(13), 1327–1329. doi.org/10.1001/jama.2015.7023

Buliung, R. N., Mitra, R., & Faulkner, G. (2009). Active school transportation in the Greater Toronto Area, Canada: An exploration of trends in space and time (1986–2006). *Preventive Medicine*, *48*(6), 507–512. doi.org/10.1016/j.ypmed.2009.03.001

Bushman, B. J., & Huesmann, L. R. (2014). Twenty-five years of research on violence in digital games and aggression revisited. *European Psychologist*, *19*(1), 47–55. doi.org/10.1027/1016-9040/a000164

Buss, D. M. (2021). *When men behave badly: The hidden roots of sexual deception, harassment, and assault*. Little, Brown Spark.

Cai, J. Y., Curchin, E., Coan, T., & Fremstad, S. (2023, March 30). *Are young men falling behind young women? The NEET rate helps shed light on the matter*. Center for Economic and Policy Research. cepr.net/report/are-young-men-falling-behind-young-women-the-neet-rate-helps-shed-light-on-the-matter/

Carhart-Harris, R. L., Erritzoe, D., Williams, T., Stone, J. M., Reed, L. J., Colasanti, A., Tyacke, R. J., Leech, R., Malizia, A. L., Murphy, K., Hobden, P., Evans, J., Feilding, A., Wise, R. G., & Nutt, D. J. (2012). Neural correlates of the psychedelic state as determined by fMRI studies with psilocybin. *Proceedings of the National Academy of Sciences*, *109*(6), 2138–2143. doi.org/10.1073/pnas.1119598109

Carr, N. (2012). *The shallows: What the internet is doing to our brains*. W. W. Norton.

Casey, M., & Nzau, S. (2019, September 11). The differing impact of automation on men and women's work. Brookings Institution. www.brookings.edu/articles/the-differing-impact-of-automation-on-men-and-womens-work/

Centers for Disease Control and Prevention. (n.d.). *WISQARS fatal and nonfatal injury reports* [Data set]. wisqars.cdc.gov/reports/

Centers for Disease Control and Prevention. (2015). *School health policies and practices study 2014*. www.cdc.gov/healthyyouth/data/shpps/pdf/shpps-508-final_101315.pdf

Chang, A.-M., Aeschbach, D., Duffy, J. F., & Czeisler, C. A. (2014). Evening use of light-emitting eReaders negatively affects sleep, circadian timing, and next-morning alertness. *Proceedings of the National Academy of Sciences of the United States of America*, *112*(4), 1232–1237. doi.org/10.1073/pnas.1418490112

Charlton, J. P., & Danforth, I. D. W. (2007). Distinguishing addiction and high engagement in the context of online game playing. *Computers in Human Behavior*, *23*(3), 1531–1548. doi.org/10.1016/j.chb.2005.07.002

Chatard, A., Bocage-Barthélémy, Y., Selimbegović, L., & Guimond, S. (2017). The woman who wasn't there: Converging evidence that subliminal social comparison affects self-evaluation. *Journal of Experimental Social Psychology*, *73*, 1–13. doi.org/10.1016/j.jesp.2017.05.005

Chen, X., Li, M., & Wei, Q. (2019). Agency and communion from the perspective of self versus others: The moderating role of social class. *Frontiers in Psychology*, *10*. doi.org/10.3389/fpsyg.2019.02867

Cohn, J. F., & Tronick, E. Z. (1987). Mother–infant face-to-face interaction: The sequence of dyadic states at 3, 6, and 9 months. *Developmental Psychology, 23*(1), 68–77. doi.org/10.1037/0012-1649.23.1.68

Coleman, E., Radix, A. E., Bouman, W. P., Brown, G. R., De Vries, A. L., Deutsch, M. B., & Arcelus, J. (2022). Standards of care for the health of transgender and gender diverse people, version 8. *International Journal of Transgender Health, 23*(supl), S1–S259. doi.org/10.1080/26895269.2022.2100644

Common Sense Media. (n.d.). *Parenting, media, and everything in between.* Common Sense Media. www.commonsensemedia.org/articles/social-media

Conner, J. O., Crawford, E., & Galioto, M. (2023). The mental health effects of student activism: Persisting despite psychological costs. *Journal of Adolescent Research, 38*(1), 80–109. doi.org/10.1177/07435584211006789

Conze, E. (1954). *Buddhist texts through the ages.* Philosophical Library.

Cosma, A., Stevens, G., Martin, G., Duinhof, E. L., Walsh, S. D., Garcia-Moya, I., Költő, A., Gobina, I., Canale, N., Catunda, C., Inchley, J., & de Looze, M. (2020). Cross-national time trends in adolescent mental well-being from 2002 to 2018 and the explanatory role of schoolwork pressure. *Journal of Adolescent Health, 66*(6S), S50–S58. doi.org/10.1016/j.jadohealth.2020.02.010

Council on Communications and Media. (2016). Media and young minds. *Pediatrics, 138*(5), Article e20162591. doi.org/10.1542/peds.2016-2591

Cox, D. A. (2021, June 29). Men's social circles are shrinking. Survey Center on American Life. www.americansurveycenter.org/why-mens-social-circles-are-shrinking/

Coyne, S. M., Hurst, J. L., Dyer, W. J., Hunt, Q., Schvaneveldt, E., Brown, S., & Jones, G. (2021). Suicide risk in emerging adulthood: Associations with screen time over 10 years. *Journal of Youth and Adolescence, 50,* 2324–2338. doi.org/10.1007/s10964-020-01389-6

Crick, N. R., & Grotpeter, J. K. (1995). Relational aggression, gender, and social-psychological adjustment. *Child Development, 66*(3), 710–722. doi.org/10.2307/1131945

Curran, T., & Hill, A. P. (2019). Perfectionism is increasing over time: A meta-analysis of birth cohort differences from 1989 to 2016. *Psychological Bulletin, 145*(4), 410–429. doi.org/10.1037/bul0000138

Cybulski, L., Ashcroft, D. M., Carr, M. J., Garg, S., Chew-Graham, C. A., Kapur, N., & Webb, R. T. (2021). Temporal trends in annual incidence rates for psychiatric disorders and self-harm among children and adolescents in the UK, 2003–2018. *BMC Psychiatry, 21*(1). doi.org/10.1186/s12888-021-03235-w

Dahl, R. E. (2008). Biological, developmental, and neurobehavioral factors relevant to adolescent driving risks. *American Journal of Preventive Medicine, 35*(3), S278–S284. doi.org/10.1016/j.amepre.2008.06.013

Damour, L. (2016). *Untangled: Guiding teenage girls through the seven transitions into adulthood.* Random House.

Darwin, C. (1998). *The descent of man and selection in relation to sex.* Original work published 1871. Amherst, N.Y.: Prometheus Books.

Davidson, R. J., & Lutz, A. (2008). Buddha's brain: Neuroplasticity and meditation. *IEEE Signal Processing Magazine, 25*(1), 176–174. doi.org/10.1109/msp.2008.4431873

Davis-berman, J., & Berman, D. S. (1989). The wilderness therapy program: An empirical study of its effects with adolescents in an outpatient setting. *Journal of Contemporary Psychotherapy, 19*(4), 271–281. doi.org/10.1007/BF00946092

Dawkins, R. (2006). *The God delusion.* Houghton Mifflin.

Dee, T. S. (2006). The why chromosome. How a teacher's gender affects boys and girls. *Education Next*, *6*(4), 68–75. eric.ed.gov/?id=EJ763353

de Graaf, N. M., Giovanardi, G., Zitz, C., & Carmichael, P. (2018). Sex ratio in children and adolescents referred to the Gender Identity Development Service in the UK (2009–2016). *Archives of Sexual Behavior*, *47*, 1301–1304. doi.org/10.1007/s10508 -018-1204-9

DeLoache, J., Chiong, C., Sherman, K., Islam, N., Vanderborght, M., Troseth, G., Strouse, G. A., & O'Doherty, K. (2010). Do babies learn from baby media? *Psychological Science*, *21*(11), 1570–1574. doi.org/10.1177/0956797610384145

Descormiers, K., & Corrado, R. R. (2016). The right to belong: Individual motives and youth gang initiation rites. *Deviant Behavior*, *37*(11), 1341–1359. doi.org/1.1080 /01639625.2016.1177390

DeSteno, D. (2021). *How God works: The science behind the benefits of religion*. Simon & Schuster.

Diaz, S., & Bailey, J. M. (2023). Rapid onset gender dysphoria: Parent reports on 1655 possible cases. *Archives of Sexual Behavior*, *52*(3), 1031–1043. doi.org/10.1007 /s10508-023-02576-9

Dodd, H. F., FitzGibbon, L., Watson, B. E., & Nesbit, R. J. (2021). Children's play and independent mobility in 2020: Results from the British children's play survey. *International Journal of Environmental Research and Public Health*, *18*(8), 4334. doi.org/10.3390/ijerph18084334

Doepke, M., Sorrenti, G., & Zilibotti, F. (2019). The economics of parenting. *Annual Review of Economics*, *11*, 55–84. doi.org/10.1146/annurev-economics-080218-030156

Donevan, M., Jonsson, L., Bladh, M., Priebe, G., Fredlund, C., & Svedin, C. G. (2022). Adolescents' use of pornography: Trends over a ten-year period in Sweden. *Archives of Sexual Behavior*, *51*, 1125–1140. doi.org/10.1007/s10508-021-02084-8

Dorahy, M. J., Brand, B. L., Şar, V., Krüger, C., Stavropoulos, P., Martínez-Taboas, A., Lewis-Fernández, R., & Middleton, W. (2014). Dissociative identity disorder: An empirical overview. *Australian and New Zealand Journal of Psychiatry*, *48*(5), 402–417. doi.org/10.1177/0004867414527523

Durkheim, É. (1951). *Suicide, a study in sociology* (J. A. Spaulding & G. Simpson, Trans.). Original work published 1897. Free Press.

Durkheim, É. (2008). *The elementary forms of religious life* (C. Cosman, Trans.). Original work published 1912. Oxford University Press.

Dwulit, A. D., & Rzymski, P. (2019). The potential associations of pornography use with sexual dysfunctions: An integrative literature review of observational studies. *Journal of Clinical Medicine*, *8*(7), 914. doi.org/10.3390/jcm8070914

Dwyer, R. J., Kushlev, K., & Dunn, E. W. (2018). Smartphone use undermines enjoyment of face-to-face social interactions. *Journal of Experimental Social Psychology*, *78*, 233–239. doi.org/10.1016/j.jesp.2017.10.007

Eagly, A. H., Nater, C., Miller, D. I., Kaufmann, M., & Sczesny, S. (2020). Gender stereotypes have changed: A cross-temporal meta-analysis of U.S. public opinion polls from 1946 to 2018. *American Psychologist*, *75*(3), 301–315. doi.org/10.1037 /amp0000494

Economides, M., Martman, J., Bell, M. J., & Sanderson, B. (2018). Improvements in stress, affect, and irritability following brief use of a mindfulness-based smartphone app: A randomized controlled trial. *Mindfulness*, *9*(5), 1584–1593. doi.org /10.1007/s12671-018-0905-4

Edmondson, A. (1999). Psychological safety and learning behavior in work teams. *Administrative Science Quarterly, 44*(2), 350–383. doi.org/10.2307/2666999

Ehrenreich, B. (2006). *Dancing in the streets: A history of collective joy.* Metropolitan Books/Henry Holt.

Eime, R. M., Young, J. A., Harvey, J. T., Charity, M. J., & Payne, W. R. (2013). A systematic review of the psychological and social benefits of participation in sport for children and adolescents: Informing development of a conceptual model of health through sport. *International Journal of Behavioral Nutrition and Physical Activity, 10*(1), Article 98. doi.org/10.1186/1479-5868-10-98

Elson, M., & Ferguson, C. J. (2014). Twenty-five years of research on violence in digital games and aggression: Empirical evidence, perspectives, and a debate gone astray. *European Psychologist, 19*(1), 33–46. doi.org/10.1027/1016-9040/a000147

Emerson, R. W. (1836). *Nature.* James Munroe. archive.vcu.edu/english/engweb/transcendentalism/authors/emerson/nature.html.

Epictetus. (1890). *The Enchiridion* (G. Long, Trans.). Original work published ca. 125 CE. George Bell and Sons.

Eyal, N. (2014). *Hooked: How to build habit-forming products.* Portfolio.

Eyal, N. (2019). *Indistractable: How to control your attention and choose your life.* Ben-Bella Books.

Fam, J. Y. (2018). Prevalence of internet gaming disorder in adolescents: A meta-analysis across three decades. *Scandinavian Journal of Psychology, 59*(5), 524–531. doi.org/10.1111/sjop.12459

Ferguson, C. J., Bean, A. M., Nielsen, R. K. L., & Smyth, M. P. (2020). Policy on unreliable game addiction diagnoses puts the cart before the horse. *Psychology of Popular Media, 9*(4), 533–540. doi.org/10.1037/ppm0000249

Ferguson, C. J., Coulson, M., & Barnett, J. (2011). A meta-analysis of pathological gaming prevalence and comorbidity with mental health, academic, and social problems. *Journal of Psychiatric Research, 45*(12), 1573–1578. doi.org/10.1016/j.jpsychires.2011.09.005

Filipe, M. G., Magalhães, S., Veloso, A. S., Costa, A. F., Ribeiro, L., Araújo, P., Castro, S. L., & Limpo, T. (2021). Exploring the effects of meditation techniques used by mindfulness-based programs on the cognitive, social-emotional, and academic skills of children: A systematic review. *Frontiers in Psychology, 12*, Article 660650. doi.org/10.3389/fpsyg.2021.660650

Finlay, B. B., & Arrieta, M.-C. (2016). *Let them eat dirt: Saving your child from an over-sanitized world.* Algonquin Books.

Fiske, S. T. (2011). *Envy up, scorn down: How status divides us.* Russell Sage Foundation.

Flanders, J. L., Leo, V., Paquette, D., Pihl, R. O., & Séguin, J. R. (2009). Rough-and-tumble play and the regulation of aggression: An observational study of father–child play dyads. *Aggressive Behavior, 35*(4), 285–295. doi.org/10.1002/ab.20309

Flynn, R. M., Shaman, N. J., & Redleaf, D. L. (2023). The unintended consequences of "lack of supervision" child neglect laws: How developmental science can inform policies about childhood independence and child protection. *Social Policy Report, 36*(1), 1–38. doi.org/10.1002/sop2.27

Food and Drug Administration. (2010, March 19). Regulations restricting the sale and distribution of cigarettes and smokeless tobacco to protect children and adolescents. *Federal Register, 75*(53), 13225–13232. www.govinfo.gov/content/pkg /FR-2010-03-19/pdf/2010-6087.pdf

Fowler, J. H., & Christakis, N. A. (2008). Dynamic spread of happiness in a large social network: Longitudinal analysis over 20 years in the Framingham Heart Study. *BMJ, 337*, Article a2338. doi.org/10.1136/bmj.a2338

Fuhrmann, D., Knoll, L. J., & Blakemore, S. (2015). Adolescence as a sensitive period of brain development. *Trends in Cognitive Sciences, 19*(10), 558–566. doi.org/10 .1016/j.tics.2015.07.008

Furedi, F. (2001). *Paranoid parenting: Abandon your anxieties and be a good parent.* Allen Lane.

Fyfe-Johnson, A. L., Hazlehurst, M. F., Perrins, S. P., Bratman, G. N., Thomas, R., Garrett, K. A., Hafferty, K. R., Cullaz, T. M., Marcuse, E. K., & Tandon, P. S. (2021). Nature and children's health: A systematic review. *Pediatrics, 148*(4), Article e2020049155. doi.org/10.1542/peds.2020-049155

Gabrielsen, L. E., Eskedal, L. T., Mesel, T., Aasen, G. O., Hirte, M., Kerlefsen, R. E., Palucha, V., & Fernee, C. R. (2019). The effectiveness of wilderness therapy as mental health treatment for adolescents in Norway: A mixed methods evaluation. *International Journal of Adolescence and Youth, 24*(3), 282–296. doi.org/10 .1080/02673843.2018.1528166

Garbarino, S., Lanteri, P., Bragazzi, N. L., Magnavita, N., & Scoditti, E. (2021). Role of sleep deprivation in immune-related disease risk and outcomes. *Communications Biology, 4*, 1304. doi.org/10.1038/s42003-021-02825-4

Garrido, E. C., Issa, T., Esteban, P. G., & Delgado, S. C. (2021). A descriptive literature review of phubbing behaviors. *Heliyon, 7*(5), Article e07037. doi.org/10.1016/j .heliyon.2021.e07037

Garriguet, D. (2021). *Portrait of youth in Canada: Data report—Chapter 1: Health of youth in Canada* (Catalogue No. 42-28-0001). Statistics Canada. www150.statcan .gc.ca/n1/en/pub/42-28-0001/2021001/article/00001-eng.pdf?st=ZQk8_2Sl

Garrison, M. M., & Christakis, D. A. (2012). The impact of a healthy media use intervention on sleep in preschool children. *Pediatrics, 130*(3), 492–499. doi.org/10 .1542/peds.2011-3153

Gemmell, E., Ramsden, R., Brussoni, M., & Brauer, M. (2023). Influence of neighborhood built environments on the outdoor free play of young children: A systematic, mixed-studies review and thematic synthesis. *Journal of Urban Health, 100*(1), 118–150. doi.org/10.1007/s11524-022-00696-6

Gillis, H. L., Speelman, E., Linville, N., Bailey, E., Kalle, A., Oglesbee, N., Sandlin, J., Thompson, L., & Jensen, J. (2016). Meta-analysis of treatment outcomes measured by the Y-OQ and Y-OQ-SR comparing wilderness and non-wilderness treatment programs. *Child and Youth Care Forum, 45*(6), 851–863. doi.org/10.1007/s10566-016-9360-3

GlobalWebIndex. (2018). *Social flagship report 2018.* www.gwi.com/hubfs/Downloads /Social-H2-2018-report.pdf

GlobalWebIndex. (2021). *Social media by generation.* 304927.fs1.hubspotusercontent -na1.net/hubfs/304927/Social%20media%20by%20generation%20-%20Global %20-%20Web_Friendly_6.pdf

Goldstone, A., Javitz, H. S., Claudatos, S. A., Buysse, D. J., Hasler, B. P., de Zambotti, M., Clark, D. B., Franzen, P. L., Prouty, D. E., Colrain, I. M., & Baker, F. C. (2020). Sleep disturbance predicts depression symptoms in early adolescence: Initial findings from the adolescent brain cognitive development study. *Journal of Adolescent Health*, *66*(5), 567–574. doi.org/10.1016/j.jadohealth.2019.12.005

Gopnik, A. (2016). *The gardener and the carpenter: What the new science of child development tells us about the relationship between parents and children.* Farrar, Straus and Giroux.

Götz, F. M., Gosling, S. D., & Rentfrow, P. J. (2022). Small effects: The indispensable foundation for a cumulative psychological science. *Perspectives on Psychological Science*, *17*(1), 205–215. doi.org/10.1177/1745691620984483

Goyal, M., Singh, S., Sibinga, E. M. S., Gould, N. F., Rowland-Seymour, A., Sharma, R., Berger, Z., Sleicher, D., Maron, D. D., Shihab, H. M., Ranasinghe, P. D., Linn, S., Saha, S., Bass, E. B., & Haythornthwaite, J. A. (2014). Meditation programs for psychological stress and well-being. *JAMA Internal Medicine*, *174*(3), 357–368. doi.org/10.1001/jamainternmed.2013.13018

Granic, I., Lobel, A., & Engels, R. C. M. E. (2014). The benefits of playing video games. *American Psychologist*, *69*(1), 66–78. doi.org/10.1037/a0034857

Grant, J. E., Potenza, M. N., Weinstein, A., & Gorelick, D. A. (2010). Introduction to behavioral addictions. *The American Journal of Drug and Alcohol Abuse*, *36*(5), 233–241. doi.org/10.3109/00952990.2010.491884

Grassini, S. (2022). A systematic review and meta-analysis of nature walk as an intervention for anxiety and depression. *Journal of Clinical Medicine*, *11*(6), 1731. doi.org/10.3390/jcm11061731

Gray, J. A. (1982). *The neuropsychology of anxiety: An enquiry into the functions of the septo-hippocampal system.* Clarendon Press/Oxford University Press.

Gray, P. (2011). The decline of play and the rise of psychopathology in children and adolescents. *American Journal of Play*, *3*(4), 443–463. www.psycnet.apa.org/record/2014-22137-001

Gray, P. (2013). The value of a play-filled childhood in development of the hunter-gatherer individual. In D. Narvaez, J. Panksepp, A. N. Schore, & T. R. Gleason (Eds.), *Evolution, early experience and human development: From research to practice and policy* (pp. 352–370). Oxford University Press.

Gray, P. (2018). Evolutionary functions of play: Practice, resilience, innovation, and cooperation. In P. K. Smith & J. L. Roopnarine (Eds.), *The Cambridge handbook of play: Developmental and disciplinary perspectives* (pp. 84–102). Cambridge University Press.

Gray, P. (2023). The special value of age-mixed play I: How age mixing promotes learning. *Play Makes Us Human.* petergray.substack.com/p/10-the-special-value-of-age-mixed

Gray, P., Lancy, D. F., & Bjorklund, D. F. (2023). Decline in independent activity as a cause of decline in children's mental wellbeing: Summary of the evidence. *Journal of Pediatrics*, *260*(2), 113352. doi.org/10.1016/j.jpeds.2023.02.004

Green, A., Cohen-Zion, M., Haim, A., & Dagan, Y. (2017). Evening light exposure to computer screens disrupts human sleep, biological rhythms, and attention abilities. *Chronobiology International*, *34*(7), 855–865. doi.org/10.1080/07420528.2017.1324878

Greitemeyer, T., & Mügge, D. O. (2014). Video games do affect social outcomes: A meta-analytic review of the effects of violent and prosocial video game play. *Personality and Social Psychology Bulletin*, *40*(5), 578–589. doi.org/10.1177/0146167213520459

Grigoriev, A. I., & Egorov, A. D. (1992). General mechanisms of the effect of weightlessness on the human body. *Advances in Space Biology and Medicine, 2,* 1–42. doi.org/10.1016/s1569-2574(08)60016-7

Guisinger, S., & Blatt, S. J. (1994). Individuality and relatedness: Evolution of a fundamental dialectic. *American Psychologist, 49*(2), 104–111. doi.org/10.1037/0003-066X .49.2.104

Guo, N., Tsun Luk, T., Wu, Y., Lai, A. Y., Li, Y., Cheung, D. Y. T., Wong, J. Y., Fong, D. Y. T., & Wang, M. P. (2022). Between- and within-person associations of mobile gaming time and total screen time with sleep problems in young adults: Daily assessment study. *Addictive Behaviors, 134,* 107408. doi.org/10.1016/j.addbeh.2022 .107408

Haapala, E. A., Väistö, J., Lintu, N., Westgate, K., Ekelund, U., Poikkeus, A.-M., Brage, S., & Lakka, T. A. (2017). Physical activity and sedentary time in relation to academic achievement in children. *Journal of Science and Medicine in Sport, 20*(6), 583–589. doi.org/10.1016/j.jsams.2016.11.003

Haidt, J. (2012). *The righteous mind: Why good people are divided by politics and religion.* Pantheon.

Haidt, J. (2023, February 23). Social media is a major cause of the mental illness epidemic in teen girls. Here's the evidence. *After Babel.* www.afterbabel.com/p/social -media-mental-illness-epidemic

Haidt, J. (2023, March 9). Why the mental health of liberal girls sank first and fastest. *After Babel.* www.afterbabel.com/p/mental-health-liberal-girls

Haidt, J. (2023, April 17). Why some researchers think I'm wrong about social media and mental illness. *After Babel.* www.afterbabel.com/p/why-some-researchers-think -im-wrong

Haidt, J., & George, E. (2023, April 12). Do the kids think they're alright? *After Babel.* www.afterbabel.com/p/do-the-kids-think-theyre-alright

Haidt, J., Park, Y. J., & Bentov, Y. (ongoing). Free play and mental health: A collaborative review. Unpublished manuscript, New York University. anxiousgeneration .com/reviews

Haidt, J., & Rausch, Z. (ongoing). Alternative hypotheses to the adolescent mental illness crisis: A collaborative review. Unpublished manuscript, New York University. anxiousgeneration.com/reviews

Haidt, J., & Rausch, Z. (ongoing). The coddling of the Canadian mind? A collaborative review. Unpublished manuscript, New York University. anxiousgeneration .com/reviews

Haidt, J., & Rausch, Z. (ongoing). The effects of phone-free schools: A collaborative review. Unpublished manuscript, New York University. anxiousgeneration.com /reviews

Haidt, J., & Rausch, Z. (ongoing). The impact of screens on infants, toddlers, and pre-schoolers: A collaborative review. Unpublished manuscript, New York University. anxiousgeneration.com/reviews

Haidt, J., Rausch, Z., & Twenge, J. (ongoing). Adolescent mood disorders since 2010: A collaborative review. Unpublished manuscript, New York University. anxiousgeneration.com/reviews

Haidt, J., Rausch, Z., & Twenge, J. (ongoing). Social media and mental health: A collaborative review. Unpublished manuscript, New York University. tinyurl.com /SocialMediaMentalHealthReview

Halldorsdottir, T., Thorisdottir, I. E., Meyers, C. C. A., Asgeirsdottir, B. B., Kristjansson,

A. L., Valdimarsdottir, H. B., Allegrante, J. P., & Sigfusdottir, I. D. (2021). Adolescent well-being amid the COVID-19 pandemic: Are girls struggling more than boys? *JCPP Advances, 1*(2), Article e12027. doi.org/10.1002/jcv2.12027

Haltigan, J. D., Pringsheim, T. M., & Rajkumar, G. (2023). Social media as an incubator of personality and behavioral psychopathology: Symptom and disorder authenticity or psychosomatic social contagion? *Comprehensive Psychiatry, 121*, Article 152362. doi.org/10.1016/j.comppsych.2022.152362

Hamilton, J. P., Farmer, M., Fogelman, P., & Gotlib, I. H. (2015). Depressive rumination, the default-mode network, and the dark matter of clinical neuroscience. *Biological Psychiatry, 78*(4), 224–230. doi.org/10.1016/j.biopsych.2015.02.020

Hamm, P. B., Billica, R. D., Johnson, G. S., Wear, M. L., & Pool, S. L. (1998, February 1). Risk of cancer mortality among the Longitudinal Study of Astronaut Health (LSAH) participants. *Aviation, Space, and Environmental Medicine, 69*(2), 142–144. pubmed.ncbi.nlm.nih.gov/9491253/

Hancock, J., Liu, S. X., Luo, M., & Mieczkowski, H. (2022). Psychological well-being and social media use: A meta-analysis of associations between social media use and depression, anxiety, loneliness, eudaimonic, hedonic, and social well-being. SSRN. dx.doi.org/10.2139/ssrn.4053961

Hari, J. (2022). *Stolen focus: Why you can't pay attention—and how to think deeply again.* Crown.

Harris, P. L. (1989). *Children and emotion: The development of psychological understanding.* Basil Blackwell.

Haslam, N. (2016). Concept creep: Psychology's expanding concepts of harm and pathology. *Psychological Inquiry, 27*(1), 1–17. doi.org/10.1080/1047840X.2016.1082418

Hassett, J. M., Siebert, E. R., & Wallen, K. (2008). Sex differences in rhesus monkey toy preferences parallel those of children. *Hormones and Behavior, 54*(3), 359–364. doi.org/10.1016/j.yhbeh.2008.03.008

Health Behaviour in School-Aged Children (HBSC). (2002–2018). *HBSC study* [Data sets]. University of Bergen. www.uib.no/en/hbscdata/113290/open-access

Hebb, D. O. (1949). *The organization of behavior: A neuropsychological theory.* Wiley.

Henrich, J. (2015). *The secret of our success: How culture is driving human evolution, domesticating our species, and making us smarter.* Princeton University Press.

Henrich, J., & Gil-White, F. J. (2001). The evolution of prestige: Freely conferred deference as a mechanism for enhancing the benefits of cultural transmission. *Evolution and Human Behavior, 22*(3), 165–196. doi.org/10.1016/s1090-5138(00)00071-4

Higher Education Research Institute (HERI). (2023). *CIRP freshman survey trends: 1966 to 2008* [Data sets]. heri.ucla.edu/data-archive/

Hillman, M., Adams, J., & Whitelegg, J. (1990). *One false move . . . : A study of children's independent mobility.* PSI.

Hisler, G., Twenge, J. M., & Krizan, Z. (2020). Associations between screen time and short sleep duration among adolescents varies by media type: Evidence from a cohort study. *Sleep Medicine, 66*, 92–102. doi.org/10.1016/j.sleep.2019.08.007

Hofferth, S. L., & Sandberg, J. F. (2001). How American children spend their time. *Journal of Marriage and Family, 63*(2), 295–308. doi.org/10.1111/j.1741-3737.2001.00295.x

Hoffmann, M. D., Barnes, J. D., Tremblay, M. S., & Guerrero, M. D. (2022). Associations between organized sport participation and mental health difficulties: Data from over 11,000 US children and adolescents. *PLoS ONE, 17*(6), Article e0268583. doi.org/10.1371/journal.pone.0268583

Howard, P. K. (2014). *The rule of nobody: Saving America from dead laws and broken government*. W. W. Norton.

Hsu, N., Badura, K. L., Newman, D. A., & Speach, M. E. P. (2021). Gender, "masculinity," and "femininity": A meta-analytic review of gender differences in agency and communion. *Psychological Bulletin, 147*(10), 987–1011. doi.org/10.1037/bul0000343

Hummer, D. L., & Lee, T. M. (2016). Daily timing of the adolescent sleep phase: Insights from a cross-species comparison. *Neuroscience and Biobehavioral Reviews, 70*, 171–181. doi.org/10.1016/j.neubiorev.2016.07.023

Hunt, M. G., Marx, R., Lipson, C., & Young, J. (2018). No more FOMO: Limiting social media decreases loneliness and depression. *Journal of Social and Clinical Psychology, 37*(10), 751–768. doi.org/10.1521/jscp.2018.37.10.751

Ishizuka, P. (2018). Social class, gender, and contemporary parenting standards in the United States: Evidence from a national survey experiment. *Social Forces, 98*(1), 31–58. doi.org/10.1093/sf/soy107

James, W. (1890). *The principles of psychology*. Classics in the History of Psychology. psychclassics.yorku.ca/James/Principles/index.htm

Jefferson, T. (1771, August 3). *From Thomas Jefferson to Robert Skipwith, with a list of books for a private library, 3 August 1771*. Founders Online, National Archives. www.founders.archives.gov/documents/Jefferson/01-01-02-0056

Johnson, J. S., & Newport, E. L. (1989). Critical period effects in second language learning: the influence of maturational state on the acquisition of English as a second language. *Cognitive Psychology, 21*(1), 60–99. doi.org/10.1016/0010-0285(89)90003-0

Joiner, R., Mizen, E., Pinnell, B., Siddique, L., Bradley, A., & Trevalyen, S. (2023). The effect of different types of TikTok dance challenge videos on young women's body satisfaction. *Computers in Human Behavior, 147*, Article 107856. doi.org/10.1016/j.chb.2023.107856

Joshi, A., & Hinkley, T. (2021, August). *Too much time on screens? Screen time effects and guidelines for children and young people*. Australian Institute of Family Studies. aifs.gov.au/resources/short-articles/too-much-time-screens

Kahlenberg, S. M., & Wrangham, R. W. (2010). Sex differences in chimpanzees' use of sticks as play objects resemble those of children. *Current Biology, 20*(24), R1067–R1068. doi.org/10.1016/j.cub.2010.11.024

Kaltiala-Heino, R., Sumia, M., Työläjärvi, M., & Lindberg, N. (2015). Two years of gender identity service for minors: overrepresentation of natal girls with severe problems in adolescent development. *Child and Adolescent Psychiatry and Mental Health, 9*(1), 1–9. doi.org/10.1186/s13034-015-0042-y

Kannan, V. D., & Veazie, P. J. (2023). US trends in social isolation, social engagement, and companionship—nationally and by age, sex, race/ethnicity, family income, and work hours, 2003–2020. *SSM–Population Health, 21*, Article 101331. doi.org/10.1016/j.ssmph.2022.101331

Kaufmann, E. (2022, May 30). Born this way? The rise of LGBT as a social and political identity. Center for the Study of Partisanship and Ideology. www.cspicenter.com/p/born-this-way-the-rise-of-lgbt-as-a-social-and-political-identity

Keeler, R. (2020). *Adventures in risky play. What is your yes?* Exchange Press.

Kelly, Y., Zilanawala, A., Booker, C., & Sacker, A. (2018). Social media use and adolescent mental health: Findings from the UK millennium cohort study. *eClinicalMedicine*, 6, 59–68. doi.org/10.1016/j.eclinm.2018.12.005

Keltner, D. (2023). *Awe: The new science of everyday wonder and how it can transform your life*. Penguin Press.

Keltner, D., & Haidt, J. (2003). Approaching awe, a moral, spiritual, and aesthetic emotion. *Cognition and Emotion*, 17(2), 297–314. doi.org/10.1080/02699930302297

Kemple, K. M., Oh, J., Kenney, E., & Smith-Bonahue, T. (2016). The power of outdoor play and play in natural environments. *Childhood Education*, 92(6), 446–454. doi.org/10.1080/00094056.2016.1251793

Keng, S.-L., Smoski, M. J., & Robins, C. J. (2011). Effects of mindfulness on psychological health: A review of empirical studies. *Clinical Psychology Review*, 31(6), 1041–1056. doi.org/10.1016/j.cpr.2011.04.006

Kennedy, R. S. (2021). Bullying trends in the United States: A meta-regression. *Trauma, Violence, and Abuse*, 22(4), 914–927. doi.org/10.1177/1524838019888555

Khan, A., Reyad, M. A. H., Edwards, E., & Horwood, S. (2023). Associations between adolescent sleep difficulties and active versus passive screen time across 38 countries. *Journal of Affective Disorders*, 320, 298–304. doi.org/10.1016/j.jad.2022.09.137

Kierkegaard, S. (2009). *Upbuilding discourses in various spirits* (H. V. Hong & E. H. Hong, Trans.). Original work published 1847. Princeton University Press.

Kim, I., Kim, R., Kim, H., Kim, D., Han, K., Lee, P. H., Mark, G., & Lee, U. (2019). Understanding smartphone usage in college classrooms: A long-term measurement study. *Computers and Education*, 141, 103611. doi.org/10.1016/j.compedu.2019.103611

Kim, S. (2023). Doing things when others do: Temporal synchrony and subjective wellbeing. *Time and Society*. doi.org/10.1177/0961463X231184099

Kim, S., Favotto, L., Halladay, J., Wang, L., Boyle, M. H., & Georgiades, K. (2020). Differential associations between passive and active forms of screen time and adolescent mood and anxiety disorders. *Social Psychiatry and Psychiatric Epidemiology*, 55(11), 1469–1478. doi.org/10.1007/s00127-020-01833-9

King, D. L., & Delfabbro, P. H. (2019). Video game monetization (e.g., "loot boxes"): A blueprint for practical social responsibility measures. *International Journal of Mental Health and Addiction*, 17, 166–179. doi.org/10.1007/s11469-018-0009-3

King, M. L., Jr. (2012). *A gift of love: Sermons from strength to love and other preachings* (Foreword by King, C. S., & Warnock, R. G.). Beacon Press.

Kirkorian, H. L., & Choi, K. (2017). Associations between toddlers' naturalistic media experience and observed learning from screens. *Infancy*, 22(2), 271–277. doi.org/10.1111/infa.12171

Klar, M., & Kasser, T. (2009). Some benefits of being an activist: Measuring activism and its role in psychological well-being. *Political Psychology*, 30(5), 755–777. doi.org/10.1111/j.1467-9221.2009.00724.x

Kleemans, M., Daalmans, S., Carbaat, I., & Anschütz, D. (2018). Picture perfect: The direct effect of manipulated Instagram photos on body image in adolescent girls. *Media Psychology*, 21(1), 93–110. doi.org/10.1080/15213269.2016.1257392

Kovess-Masfety, V., Keyes, K., Hamilton, A., Hanson, G., Bitfoi, A., Golitz, D., Koç, C., Kuijpers, R., Lesinskiene, S., Mihova, Z., Otten, R., Fermanian, C., & Pez, O. (2016). Is time spent playing video games associated with mental health, cognitive, and social skills in young children? *Social Psychiatry and Psychiatric Epidemiology*, 51, 349–357. doi.org/10.1007/s00127-016-1179-6

Kowert, R., & Oldmeadow, J. A. (2015). Playing for social comfort: Online video game play as a social accommodator for the insecurely attached. *Computers in Human Behavior, 53*, 556–566. doi.org/10.1016/j.chb.2014.05.004

Kristensen, J. H., Pallesen, S., King, D. L., Hysing, M., & Erevik, E. K. (2021). Problematic gaming and sleep: A systematic review and meta-analysis. *Frontiers in Psychiatry, 12*. doi.org/10.3389/fpsyt.2021.675237

Lacey, T. J. (2006). *The Blackfeet*. Chelsea House.

Lange, B. P., Wühr, P., & Schwarz, S. (2021). Of time gals and mega men: Empirical findings on gender differences in digital game genre preferences and the accuracy of respective gender stereotypes. *Frontiers in Psychology, 12*, Article 657430. doi.org/10.3389/fpsyg.2021.657430

Lareau, A. (2003). *Unequal childhoods: Class, race, and family life*. University of California Press.

Latane, B., & Darley, J. M. (1968). Group inhibition of bystander intervention in emergencies. *Journal of Personality and Social Psychology, 10*(3), 215–221. doi.org/10.1037/h0026570

Latkin, C., Dayton, L., Scherkoske, M., Countess, K., & Thrul, J. (2022). What predicts climate change activism? An examination of how depressive symptoms, climate change distress, and social norms are associated with climate change activism. *Journal of Climate Change and Health, 8*, Article 100146. doi.org/10.1016/j.joclim.2022.100146

Lauricella, A. R., Cingel, D. P., Beaudoin-Ryan, L., Robb, M. B., Saphir, M., & Wartella, E. A. (2016). *The Common Sense census: Plugged-in parents of tweens and teens*. Common Sense Media.

Leary, M. R. (2005). Sociometer theory and the pursuit of relational value: Getting to the root of self-esteem. *European Review of Social Psychology, 16*, 75–111. doi.org/10.1080/10463280540000007

LeDoux, J. (1996). *The emotional brain: The mysterious underpinnings of emotional life*. Simon & Schuster.

Lee, J., Tsunetsugu, Y., Takayama, N., Park, B.-J., Li, Q., Song, C., Komatsu, M., Ikei, H., Tyrväinen, L., Kagawa, T., & Miyazaki, Y. (2014). Influence of forest therapy on cardiovascular relaxation in young adults. *Evidence-Based Complementary and Alternative Medicine, 2014*, Article ID 834360. doi.org/10.1155/2014/834360

Lembke, A. (2021). *Dopamine nation: Finding balance in the age of indulgence*. Dutton.

Lemmens, J. S., Valkenburg, P. M., & Peter, J. (2009). Development and validation of a game addiction scale for adolescents. *Media Psychology, 12*(1), 77–95. doi.org/10.1080/15213260802669458

Lenhart, A. (2012, March 12). Teens, smartphones & texting. Pew Research Center. www.pewresearch.org/internet/2012/03/19/cell-phone-ownership/

Lenhart, A. (2015, April 9). Teen, social media, and technology overview 2015: Smartphones facilitate shifts in communication landscape for teens. Pew Research Center. www.pewresearch.org/internet/2015/04/09/teens-social-media-technology-2015/

Lester, D. (1993). The effect of war on suicide rates. *European Archives of Psychiatry and Clinical Neuroscience, 242*(4), 248–249. doi.org/10.1007/bf02189971

Li, R., Lian, Q., Su, Q., Li, L., Xie, M., & Hu, J. (2020). Trends and sex disparities in school bullying victimization among U.S. youth, 2011–2019. *BMC Public Health, 20*(1), Article 1583. doi.org/10.1186/s12889-020-09677-3

Lieber, R. (2015). *The opposite of spoiled: Raising kids who are grounded, generous, and smart about money*. HarperCollins.

Littman, L. (2018). Rapid-onset gender dysphoria in adolescents and young adults: A study of parental reports. *PLoS ONE, 13*(8), e0202330. doi.org/10.1371/journal.pone.0202330

Liu, H., Chen, X., Huang, M., Yu, X., Gan, Y., Wang, J., Chen, Q., Nie, Z., & Ge, H. (2023). Screen time and childhood attention deficit hyperactivity disorder: A meta-analysis. *Reviews on Environmental Health.* doi.org/10.1515/reveh-2022-0262

Lopes, L. S., Valentini, J. P., Monteiro, T. H., Costacurta, M. C. de F., Soares, L. O. N., Telfar-Barnard, L., & Nunes, P. V. (2022). Problematic social media use and its relationship with depression or anxiety: A systematic review. *Cyberpsychology, Behavior, and Social Networking, 25*(11), 691–702. doi.org/10.1089/cyber.2021.0300

Lowe, C. J., Safati, A., & Hall, P. A. (2017). The neurocognitive consequences of sleep restriction: A meta-analytic review. *Neuroscience and Biobehavioral Reviews, 80*, 586–604. doi.org/10.1016/j.neubiorev.2017.07.010

Lucas, K., & Sherry, J. L. (2004). Sex differences in video game play: A communication-based explanation. *Communication Research, 31*(5), 499–523. doi.org/10.1177/0093650204267930

Lukianoff, G., & Haidt, J. (2018). *The coddling of the American mind: How good intentions and bad ideas are setting up a generation for failure.* Penguin Books.

Luo, Y., Moosbrugger, M., Smith, D. M., France, T. J., Ma, J., & Xiao, J. (2022). Is increased video game participation associated with reduced sense of loneliness? A systematic review and meta-analysis. *Frontiers in Public Health, 10.* www.frontiersin.org/articles/10.3389/fpubh.2022.898338

Maccoby, E. E., & Jacklin, C. N. (1974). *The psychology of sex differences.* Stanford University Press.

Madore, K. P., & Wagner, A. D. (2019). Multicosts of multitasking. *Cerebrum, 2019* (March–April), cer-04-19. www.ncbi.nlm.nih.gov/pmc/articles/PMC7075496/

Maezumi, T., & Cook, F. D. (2007). The eight awarenesses of the enlightened person: Dogen Zenji's Hachidainingaku. In T. Maezumi & B. Glassman (Eds.), *The hazy moon of enlightenment.* Wisdom Publications.

Mandryk, R. L., Frommel, J., Armstrong, A., & Johnson, D. (2020). How passion for playing World of Warcraft predicts in-game social capital, loneliness, and well-being. *Frontiers in Psychology, 11*, Article 2165. doi.org/10.3389/fpsyg.2020.02165

Männikkö, N., Ruotsalainen, H., Miettunen, J., Pontes, H. M., & Kääriäinen, M. (2020). Problematic gaming behaviour and health-related outcomes: A systematic review and meta-analysis. *Journal of Health Psychology, 25*(1), 67–81. doi.org/10.1177/1359105317740414

Marano, H. E. (2008). *A nation of wimps: The high cost of invasive parenting.* Crown Archetype.

Marchiano, L. (2017). Outbreak: on transgender teens and psychic epidemics. *Psychological Perspectives, 60*(3), 345–366. doi.org/10.1080/00332925.2017.1350804

Marcus Aurelius. (2002). *Meditations* (G. Hays, Trans.). Original work published 161–180 CE. Random House.

Markey, P. M., & Ferguson, C. J. (2017). *Moral combat: Why the war on violent video games is wrong.* BenBella Books.

Markstrom, C. A. (2008). *Empowerment of North American Indian girls: Ritual expressions at puberty.* University of Nebraska Press.

Maza, M. T., Fox, K. A., Kwon, S., Flannery, J. E., Lindquist, K. A., Prinstein, M. J., &

Telzer, E. H. (2023). Association of habitual checking behaviors on social media with longitudinal functional brain development. *JAMA Pediatrics*, *177*(2), 160–167. doi.org/10.1001/jamapediatrics.2022.4924

McCabe, B. J. (2019). Visual imprinting in birds: Behavior, models, and neural mechanisms. *Frontiers in Physiology*, *10*. doi.org/10.3389/fphys.2019.00658

McLeod, B. D., Wood, J. J., & Weisz, J. R. (2006). Examining the association between parenting and childhood anxiety: A meta-analysis. *Clinical Psychology Review*, *27*(2), 155–172. doi.org/10.1016/j.cpr.2006.09.002

McNeill, W. H. (1995). *Keeping together in time: Dance and drill in human history*. Harvard University Press.

Mercado, M. C., Holland, K. M., Leemis, R. W., Stone, D. L., & Wang, J. (2017). Trends in emergency department visits for nonfatal self-inflicted injuries among youth aged 10 to 24 years in the United States, 2001–2015. *JAMA*, *318*(19), 1931–1933. doi.org/10.1001/jama.2017.13317

Milder, C. M., Elgart, S. R., Chappell, L., Charvat, J. M., Van Baalen, M., Huff, J. L., & Semones, E. J. (2017, January 23). Cancer risk in astronauts: A constellation of uncommon consequences. *NASA Technical Reports Server (NTRS)*. ntrs.nasa.gov/citations/20160014586

Mindell, J. A., Sedmak, R., Boyle, J. T., Butler, R., & Williamson, A. A. (2016). Sleep well! A pilot study of an education campaign to improve sleep of socioeconomically disadvantaged children. *Journal of Clinical Sleep Medicine*, *12*(12), 1593–1599. jcsm.aasm.org/doi/10.5664/jcsm.6338

Ministry of Health, Labor, and Welfare. (2003, July 28). 「ひきこもり」対応ガイドライン（最終版）の作成・通知について [Creation and notification of the final version of the "Hikikomori" response guidelines]. www.mhlw.go.jp/topics/2003/07/tp0728-1.html

Minoura, Y. (1992). A sensitive period for the incorporation of a cultural meaning system: A study of Japanese children growing up in the United States. *Ethos*, *20*(3), 304–339. doi.org/10.1525/eth.1992.20.3.02a00030

Mitra, P., & Jain, A. (2023). Dissociative identity disorder. In *StatPearls [Internet]*. StatPearls. www.ncbi.nlm.nih.gov/books/NBK568768/

Monroy, M., & Keltner, D. (2023). Awe as a pathway to mental and physical health. *Perspectives on Psychological Science*, *18*(2), 309–320. doi.org/10.1177/17456916221094856

Mullan, K. (2018). Technology and children's screen-based activities in the UK: The story of the millennium so far. *Child Indicators Research*, *11*(6), 1781–1800. doi.org/10.1007/s12187-017-9509-0

Mullan, K. (2019). A child's day: Trends in time use in the UK from 1975 to 2015. *British Journal of Sociology*, *70*(3), 997–1024. doi.org/10.1111/1468-4446.12369

Müller-Vahl, K. R., Pisarenko, A., Jakubovski, E., & Fremer, C. (2022). Stop that! It's not Tourette's but a new type of mass sociogenic illness. *Brain*, *145*(2), 476–480. doi.org/10.1093/brain/awab316

Mullola, S., Ravaja, N., Lipsanen, J., Alatupa, S., Hintsanen, M., Jokela, M., & Keltikangas-Järvinen, L. (2012). Gender differences in teachers' perceptions of students' temperament, educational competence, and teachability. *British Journal of Educational Psychology*, *82*(2), 185–206. doi.org/10.1111/j.2044-8279.2010.02017.x

Murray, R., & Ramstetter, C. (2013). The crucial role of recess in school. *Pediatrics*, *131*(1), 183–188. doi.org/10.1542/peds.2012-2993

Myers, L. J., LeWitt, R. B., Gallo, R. E., & Maselli, N. M. (2017). Baby FaceTime: Can toddlers learn from online video chat? *Developmental Science, 20*(4), Article e12430. doi.org/10.1111/desc.12430

Nagata, J. M., Cortez, C. A., Dooley, E. E., Bibbins-Domingo, K., Baker, F. C., & Gabriel, K. P. (2022). Screen time and moderate-to-vigorous intensity physical activity among adolescents during the COVID-19 pandemic: Findings from the Adolescent Brain Cognitive Development Study. *Journal of Adolescent Health, 70*(4), S6. doi.org/10.1016/j.jadohealth.2022.01.014

Nagata, J. M., Ganson, K. T., Iyer, P., Chu, J., Baker, F. C., Pettee Gabriel, K., Garber, A. K., Murray, S. B., & Bibbins-Domingo, K. (2022). Sociodemographic correlates of contemporary screen time use among 9- and 10-year-old children. *Journal of Pediatrics, 240*, 213–220.e2. doi.org/10.1016/j.jpeds.2021.08.077

Nagata, J. M., Lee, C. M., Yang, J., Al-Shoaibi, A. A. A., Ganson, K. T., Testa, A., & Jackson, D. B. (2023). Associations between sexual orientation and early adolescent screen use: Findings from the Adolescent Brain Cognitive Development (ABCD) Study. *Annals of Epidemiology, 82*, 54–58.e1. doi.org/10.1016/j.annepidem.2023.03.004

Nagata, J. M., Singh, G., Sajjad, O. M., Ganson, K. T., Testa, A., Jackson, D. B., Assari, S., Murray, S. B., Bibbins-Domingo, K., & Baker, F. C. (2022). Social epidemiology of early adolescent problematic screen use in the United States. *Pediatric Research, 92*(5), 1443–1449. doi.org/10.1038/s41390-022-02176-8

National Addiction & HIV Data Archive Program. (n.d.-a). Monitoring the future: A continuing study of American youth [8th- and 10th-grade data sets]. www.icpsr.umich.edu/web/NAHDAP/series/35

National Addiction & HIV Data Archive Program. (n.d.-b). Monitoring the future: A continuing study of American youth [12th-grade data sets]. www.icpsr.umich.edu/web/NAHDAP/series/35/

National Center for Education Statistics. (n.d.). National Assessment of Educational Progress (NAEP) [Data sets]. U.S. Department of Education. www.nationsreportcard.gov/ndecore/xplore/ltt

Nauta, J., Martin-Diener, E., Martin, B. W., van Mechelen, W., & Verhagen, E. (2014). Injury risk during different physical activity behaviours in children: A systematic review with bias assessment. *Sports Medicine, 45*, 327–336. doi.org/10.1007/s40279-014-0289-0

Nesi, J., Mann, S., & Robb, M. B. (2023). *Teens and mental health: How girls really feel about social media*. Common Sense. www.commonsensemedia.org/sites/default/files/research/report/how-girls-really-feel-about-social-media-researchreport_web_final_2.pdf

New revised standard version Bible. (1989). National Council of the Churches of Christ in the U.S.A.

Nuwer, H. (1999). *Wrongs of passage: Fraternities, sororities, hazing, and binge drinking*. Indiana University Press.

O'Brien, J., & Smith, J. (2002). Childhood transformed? Risk perceptions and the decline of free play. *British Journal of Occupational Therapy, 65*(3), 123–128. doi.org/10.1177/030802260206500304

Office for National Statistics. (2022, February 24). *Young people not in education, employment, or training (NEET), UK: February 2022*. www.ons.gov.uk/employmentandlabourmarket/peoplenotinwork/unemployment/bulletins/youngpeoplenotineducationemploymentortrainingneet/february2022

Ogas, O., & Gaddam, S. (2011). *A billion wicked thoughts: What the world's largest experiment reveals about human desire.* Dutton.

Orben, A. (2020). Teenagers, screens, and social media: A narrative review of reviews and key studies. *Social Psychiatry and Psychiatric Epidemiology, 55,* 407–414. doi.org/10.1007/s00127-019-01825-4

Orben, A., & Przybylski, A. K. (2019). The association between adolescent well-being and digital technology use. *Nature Human Behaviour, 3,* 173–182. doi.org/10.1038/s41562-018-0506-1

Orben, A., Przybylski, A. K., Blakemore, S., & Kievit, R. A. (2022). Windows of developmental sensitivity to social media. *Nature Communications, 13,* Article 1649. doi.org/10.1038/s41467-022-29296-3

Orces, C. H., & Orces, J. (2020) Trends in the U.S. childhood emergency department visits for fall-related fractures, 2001–2015. *Cureus, 12*(11), Article e11629.

Organization for Economic Cooperation and Development (OECD). *PISA survey* [Data sets]. www.oecd.org/pisa/data/

Owens, J., Au, R., Carskadon, M., Millman, R., Wolfson, A., Braverman, P. K., Adelman, W. P., Breuner, C. C., Levine, D. A., Marcell, A. V., Murray, P. J., & O'Brien, R. F. (2014). Insufficient sleep in adolescents and young adults: An update on causes and consequences. *Pediatrics, 134*(3), e921–e932. dx.doi.org/10.1542/peds.2014-1696

Pallavicini, F., Pepe, A., & Mantovani, F. (2022). The effects of playing video games on stress, anxiety, depression, loneliness, and gaming disorder during the early stages of the COVID-19 pandemic: PRISMA systematic review. *Cyberpsychology, Behavior, and Social Networking, 25*(6), 334–354. doi.org/10.1089/cyber.2021.0252

Parker, K. (2021, November 8). Why the gap between men and women finishing college is growing. Pew Research Center. www.pewresearch.org/short-reads/2021/11/08/whats-behind-the-growing-gap-between-men-and-women-in-college-completion/

Parker, K., & Igielnik, R. (2020, May 14). On the cusp of adulthood and facing an uncertain future: What we know about Gen Z so far. Pew Research Center. www.pewresearch.org/social-trends/2020/05/14/on-the-cusp-of-adulthood-and-facing-an-uncertain-future-what-we-know-about-gen-z-so-far-2/

Parodi, K. B., Holt, M. K., Green, J. G., Porche, M. V., Koenig, B., & Xuan, Z. (2022). Time trends and disparities in anxiety among adolescents, 2012–2018. *Social Psychiatry and Psychiatric Epidemiology, 57*(1), 127–137. doi.org/10.1007/s00127-021-02122-9

Partelow, L. (2019). *What to make of declining enrollment in teacher preparation programs.* Center for American Progress. www.americanprogress.org/wp-content/uploads/sites/2/2019/11/TeacherPrep-report1.pdf

Paruthi, S., Brooks, L. J., D'Ambrosio, C., Hall, W. A., Kotagal, S., Lloyd, R. M., Malow, B. A., Maski, K., Nichols, C., Quan, S. F., Rosen, C. L., Troester, M. M., & Wise, M. S. (2016). Recommended amount of sleep for pediatric populations: A consensus statement of the American Academy of Sleep Medicine. *Journal of Clinical Sleep Medicine, 12*(6), 785–786. doi.org/10.5664/jcsm.5866

Pascal, B. (1966). *Pensées.* Penguin Books.

Pedersen, J. (2022). *Recreational screen media use and its effect on physical activity, sleep, and mental health in families with children.* University of Southern Denmark. doi.org/10.21996/dn60-bh82

Peracchia, S., & Curcio, G. (2018). Exposure to video games: Effects on sleep and on post-sleep cognitive abilities: A systematic review of experimental evidences. *Sleep Science*, *11*(4), 302–314. dx.doi.org/10.5935/1984-0063.20180046

Perez-Lloret, S., Videla, A. J., Richaudeau, A., Vigo, D., Rossi, M., Cardinali, D. P., & Perez-Chada, D. (2013). A multi-step pathway connecting short sleep duration to daytime somnolence, reduced attention, and poor academic performance: An exploratory cross-sectional study in teenagers. *Journal of Clinical Sleep Medicine*, *9*(5), 469–473. doi.org/10.5664/jcsm.2668

Perrault, A. A., Bayer, L., Peuvrier, M., Afyouni, A., Ghisletta, P., Brockmann, C., Spiridon, M., Vesely, S. H., Haller, D. M., Pichon, S., Perrig, S., Schwartz, S., & Sterpenich, V. (2019). Reducing the use of screen electronic devices in the evening is associated with improved sleep and daytime vigilance in adolescents. *Sleep*, *42*(9), zsz125. doi.org/10.1093/sleep/zsz125

Perrin, A., & Atske, S. (2021, March 26). About three-in-ten U.S. adults say they are "almost constantly" online. Pew Research Center. www.pewresearch.org/short-reads/2021/03/26/about-three-in-ten-u-s-adults-say-they-are-almost-constantly-online/

Pew Research Center. (2015, December 17). *Parenting in America: Outlook, worries, aspirations are strongly linked to financial situation.* www.pewresearch.org/social-trends/wp-content/uploads/sites/3/2015/12/2015-12-17_parenting-in-america_FINAL.pdf

Pew Research Center. (2019, October). *Majority of Americans say parents are doing too much for their young adult children.* www.pewresearch.org/social-trends/2019/10/23/majority-of-americans-say-parents-are-doing-too-much-for-their-young-adult-children/

Pew Research Center. (2020, July). *Parenting children in the age of screens.* www.pewresearch.org/internet/2020/07/28/parenting-children-in-the-age-of-screens/

Pew Research Center. (2021, April 7). *Internet/broadband fact sheet.* www.pewresearch.org/internet/fact-sheet/internet-broadband/

Phan, M., Jardina, J. R., Hoyle, W. S., & Chaparro, B. S. (2012). Examining the role of gender in video game usage, preference, and behavior. *Proceedings of the Human Factors and Ergonomics Society Annual Meeting*, *56*(1), 1496–1500. doi.org/10.1177/1071181312561297

Phelan, T. W. (2010). *1-2-3 magic: Effective discipline for children 2–12.* Parentmagic.

Pinker, S. (2011). *The better angels of our nature: Why violence has declined.* Viking.

Pizzol, D., Bertoldo, A., & Foresta, C. (2016). Adolescents and web porn: A new era of sexuality. *International Journal of Adolescent Medicine and Health*, *28*(2), 169–173. doi.org/10.1515/ijamh-2015-0003

Pluhar, E., McCracken, C., Griffith, K. L., Christino, M. A., Sugimoto, D., & Meehan, W. P., III. (2019). Team sport athletes may be less likely to suffer anxiety or depression than individual sport athletes. *Journal of Sports Science and Medicine*, *18*(3), 490–496.

Ponti, M., Bélanger, S., Grimes, R., Heard, J., Johnson, M., Moreau, E., Norris, M., Shaw, A., Stanwick, R., Van Lankveld, J., & Williams, R. (2017). Screen time and young children: Promoting health and development in a digital world. *Paediatrics and Child Health*, *22*(8), 461–468. doi.org/10.1093/pch/pxx123

Poulton, R., & Menzies, R. G. (2002a). Non-associative fear acquisition: A review of the evidence from retrospective and longitudinal research. *Behaviour Research and Therapy*, *40*(2), 127–149. doi.org/10.1016/s0005-7967(01)00045-6

Poulton, R., & Menzies, R. G. (2002b). Fears born and bred: Toward a more inclusive

theory of fear acquisition. *Behaviour Research and Therapy, 40*(2), 197–208. doi
.org/10.1016/s0005-7967(01)00052-3

Prescott, A. T., Sargent, J. D., & Hull, J. G. (2018). Metaanalysis of the relationship between violent video game play and physical aggression over time. *Proceedings
of the National Academy of Sciences, 115*(40), 9882–9888. doi.org/10.1073/pnas
.1611617114

Price-Feeney, M., Green, A. E., & Dorison, S. (2020). Understanding the mental health
of transgender and nonbinary youth. *Journal of Adolescent Health, 66*(6), 684–690.
doi.org/10.1016/j.jadohealth.2019.11.314

Primack, B. A., Shensa, A., Sidani, J. E., Escobar-Viera, C. G., & Fine, M. J. (2021). Temporal associations between social media use and depression. *American Journal
of Preventive Medicine, 60*(2), 179–188. doi.org/10.1016/j.amepre.2020.09.014

Przybylski, A. K. (2019). Digital screen time and pediatric sleep: Evidence from a
preregistered cohort study. *Journal of Pediatrics, 205*, 218–223.e1. doi.org/10.1016
/j.jpeds.2018.09.054

Przybylski, A. K., & Weinstein, N. (2013). Can you connect with me now? How the
presence of mobile communication technology influences face-to-face conversation quality. *Journal of Social and Personal Relationships, 30*(3), 237–246. doi.org
/10.1177/0265407512453827

Pulkki-Råback, L., Barnes, J. D., Elovainio, M., Hakulinen, C., Sourander, A., Tremblay, M. S., & Guerrero, M. D. (2022). Parental psychological problems were associated with higher screen time and the use of mature-rated media in children.
Acta Paediatrica, 111(4), 825–833. doi.org/10.1111/apa.16253

Putnam, R. D. (2000). *Bowling alone: The collapse and revival of American community.*
Simon & Schuster.

Ramey, G., & Ramey, V. A. (2010). The rug rat race. *Brookings Papers on Economic Activity, 41*, 129–199. www.brookings.edu/wp-content/uploads/2010/03/2010a_bpea
_ramey.pdf

Rasmussen, M. G. B., Pedersen, J., Olesen, L. G., Brage, S., Klakk, H., Kristensen, P. L.,
Brønd, J. C., & Grøntved, A. (2020). Short-term efficacy of reducing screen media
use on physical activity, sleep, and physiological stress in families with children
aged 4–14: Study protocol for the SCREENS randomized controlled trial. *BMC
Public Health, 20*, 380. doi.org/10.1186/s12889-020-8458-6

Raudino, A., Fergusson, D. M., & Horwood, L. J. (2013). The quality of parent/child
relationships in adolescence is associated with poor adult psychosocial adjustment.
Journal of Adolescence, 36(2), 331–340. doi.org/10.1016/j.adolescence.2012.12.002

Rausch, Z., Carlton, C., & Haidt, J. (ongoing). Social media reforms: A collaborative review. Unpublished manuscript, docs.google.com/document/d/1ULUWW1roAR3b
_EtC98eZUxYu69K_cpW5j0JsJUWXgHM/edit?usp=sharing

Rausch, Z., & Haidt, J. (2023, March 29). The teen mental illness epidemic is international, part 1: The Anglosphere. *After Babel.* www.afterbabel.com/p/international
-mental-illness-part-one

Rausch, Z., & Haidt, J. (2023, April 19). The teen mental illness epidemic is international, part 2: The Nordic nations. *After Babel.* www.afterbabel.com/p/international
-mental-illness-part-two

Rausch, Z., & Haidt, J. (2023, October 30). Suicide rates are up for Gen Z across the
anglosphere, especially for girls. *After Babel.* www.afterbabel.com/p/anglo-teen
-suicide

Rausch, Z., & Haidt, J. (2023, November). Solving the social dilemma: Many paths to reform. *After Babel.* www.afterbabel.com/p/solving-the-social-dilemma

Reed, P. (2023). Impact of social media use on executive function. *Computers in Human Behavior, 141,* Article 107598. doi.org/10.1016/j.chb.2022.107598

Reeves, R. (2022). *Of boys and men: Why the modern male is struggling, why it matters, and what to do about it.* Brookings Institution Press.

Reeves, R. (2022, September 25). Men can HEAL. *Of Boys and Men.* ofboysandmen .substack.com/p/men-can-heal

Reeves, R. (2023, March 13). The underreported rise in male suicide. *Of Boys and Men.* ofboysandmen.substack.com/p/the-underreported-rise-in-male-suicide

Reeves, R., Buckner, E., & Smith, E. (2021, January 12). The unreported gender gap in high school graduation rates. Brookings. www.brookings.edu/articles/the-unreported -gender-gap-in-high-school-graduation-rates/

Reeves, R., & Smith, E. (2020, October 7). Americans are more worried about their sons than their daughters. Brookings. www.brookings.edu/articles/americans-are -more-worried-about-their-sons-than-their-daughters/

Reeves, R., & Smith, E. (2021, October 8). The male college crisis is not just in enrollment, but completion. Brookings. www.brookings.edu/articles/the-male-college-crisis -is-not-just-in-enrollment-but-completion/

Richerson, P. J., & Boyd, R. (2004). *Not by genes alone: How culture transformed human evolution.* University of Chicago Press.

Rideout, V. (2021). *The Common Sense census: Media use by tweens and teens in America, a Common Sense Media research study, 2015.* ICPSR. doi.org/10.3886/ICPSR38018.v1

Rideout, V., Lauricella, A., & Wartella, E. (2011). *State of the science conference report: A roadmap for research on biological markers of the social environment.* Center on Social Disparities and Health, Institute for Policy Research, Northwestern University. cmhd.northwestern.edu/wp-content/uploads/2011/06/SOCconfReportSingleFinal -1.pdf

Rideout, V., Peebles, A., Mann, S., & Robb, M. B. (2022). *Common Sense census: Media use by tweens and teens, 2021.* Common Sense. www.commonsensemedia.org/sites /default/files/research/report/8-18-census-integrated-report-final-web_0.pdf

Rideout, V., & Robb, M. B. (2019). *The Common Sense census: Media use by tweens and teens, 2019.* Common Sense Media. www.commonsensemedia.org/sites/default /files/research/report/2019-census-8-to-18-full-report-updated.pdf

Roebuck, V. J. (Trans.). (2010). *The Dhammapada.* Penguin UK.

Rojcewicz, S. (1971). War and suicide. *Suicide and Life Threatening Behavior, 1*(1), 46–54. onlinelibrary.wiley.com/doi/abs/10.1111/j.1943-278X.1971.tb00598.x

Roseberry, S., Hirsh-Pasek, K., & Golinkoff, R. M. (2014). Skype me! Socially contingent interactions help toddlers learn language. *Child Development, 85*(3), 956–970. doi.org/10.1111/cdev.12166

Rosenquist, J. N., Fowler, J. H., & Christakis, N. A. (2011). Social network determinants of depression. *Molecular Psychiatry, 16,* 273–281. doi.org/10.1038/mp .2010.13

Roser, M., Ritchie, H., & Dadonaite, B. (2019). *Child and infant mortality.* Our World in Data. ourworldindata.org/child-mortality

Rosin, H. (2012). *The end of men: And the rise of women.* Riverhead Books.

Royal Society for Public Health. (2017). *Status of mind: Social media and young peo-*

ple's mental health and wellbeing. www.rsph.org.uk/static/uploaded/d125b27c-0b62
-41c5-a2c0155a8887cd01.pdf

Ruiz Pardo, A. C., & Minda, J. P. (2022). Reexamining the "brain drain" effect: A repli-
cation of Ward et al. (2017). *Acta Psychologica, 230,* 103717. doi.org/10.1016/j.actpsy
.2022.103717

Russoniello, C. V., Fish, M., & O'Brien, K. (2013). The efficacy of casual videogame play
in reducing clinical depression: A randomized controlled study. *Games for Health
Journal, 2*(6), 341–346. doi.org/10.1089/g4h.2013.0010

Sales, N. J. (2016). *American girls: Social media and the secret lives of teenagers.* Knopf.

Sampalo, M., Lázaro, E., & Luna, P.-M. (2023). Action video gaming and attention in
young adults: A systematic review. *Journal of Attention Disorders, 27*(5), 530–538.
doi.org/10.1177/10870547231153878

Sandseter, E. B. H., & Kennair, L. E. O. (2011). Children's risky play from an evolutionary
perspective: The anti-phobic effects of thrilling experiences. *Evolutionary Psychology, 9*(2), 257–284. doi.org/10.1177/147470491100900212

Sandseter, E. B. H., Kleppe, R., & Kennair, L. E. O. (2023). Risky play in children's emotion regulation, social functioning, and physical health: An evolutionary approach.
International Journal of Play, 12(1), 127–139. doi.org/10.1080/21594937.2022.2152531

Sandseter, E. B. H., Kleppe, R., & Sando, O. J. (2021). The Prevalence of Risky Play in
Young Children's Indoor and Outdoor Free Play. *Early Childhood Education Journal,
49*(2), 303–312. doi.org/10.1007/s10643-020-01074-0

Santos, R. M. S., Mendes, C. G., Miranda, D. M., & Romano-Silva, M. A. (2022). The association between screen time and attention in children: A systematic review. *Developmental Neuropsychology, 47*(4), 175–192. doi.org/10.1080/87565641.2022.2064863

Sapien Labs. (2023, May 14). *Age of first smartphone/tablet and mental wellbeing
outcomes.* sapienlabs.org/wp-content/uploads/2023/05/Sapien-Labs-Age-of-First
-Smartphone-and-Mental-Wellbeing-Outcomes.pdf

Scarr, S. (1992). Developmental theories for the 1990s: Development and individual
differences. *Child Development, 63,* 1–19.

Schneider, S. K., O'Donnell, L., & Smith, E. (2015). Trends in cyberbullying and school
bullying victimization in a regional census of high school students, 2006–2012.
Journal of School Health, 85(9), 611–620. doi.org/10.1111/josh.12290

Sewall, C. J. R., Bear, T. M., Merranko, J., & Rosen, D. (2020). How psychosocial well-
being and usage amount predict inaccuracies in retrospective estimates of digital technology use. *Mobile Media and Communication, 8*(3), 379–399. doi.org
/10.1177/2050157920902830

Shakya, H. B., & Christakis, N. A. (2017). Association of Facebook use with compro-
mised well-being: A longitudinal study. *American Journal of Epidemiology, 185*(3),
203–211. doi.org/10.1093/aje/kww189

Shaw, B., Bicket, M., Elliott, B., Fagan-Watson, B., Mocca, E., & Hillman, M. (2015).
*Children's independent mobility: An international comparison and recommenda-
tions for action.* Policy Studies Institute. www.nuffieldfoundation.org/sites/default
/files/files/7350_PSI_Report_CIM_final.pdf

Sherman, G. D., Haidt, J., & Coan, J. (2009). Viewing cute images increases behavioral
carefulness. *Emotion, 9*(2), 282–286. doi.org/10.1037/a0014904

Singh, A., Uijtdewilligen, L., Twisk, J. W. R., van Mechelen, W., & Chinapaw, M. J. M.
(2012). Physical activity and performance at school: A systematic review of the

literature including a methodological quality assessment. *Archives of Pediatrics & Adolescent Medicine*, *166*(1), 49–55. doi.org/10.1001/archpediatrics.2011.716

Shoebridge, P., & Gowers, S. (2000). Parental high concern and adolescent-onset anorexia nervosa: A case-control study to investigate direction of causality. *British Journal of Psychiatry*, *176*(2), 132–137. doi.org/10.1192/bjp.176.2.132

Skenazy, L. (2009). *Free-range kids*. Jossey-Bass.

Skowronek, J., Seifert, A., & Lindberg, S. (2023). The mere presence of a smartphone reduces basal attentional performance. *Scientific Reports*, *13*(1), 9363. doi.org/10.1038/s41598-023-36256-4

Snodgrass, J. G., Lacy, M. G., & Cole, S. W. (2022). Internet gaming, embodied distress, and psychosocial well-being: A syndemic-syndaimonic continuum. *Social Science and Medicine*, *295*, Article 112728. doi.org/10.1016/j.socscimed.2019.112728

Statista Research Department. (2023, June 2). Post-baccalaureate enrollment numbers U.S. 1976–2030, by gender. Statista. www.statista.com/statistics/236654/us-post-baccalaureate-enrollment-by-gender/

Stein, D. (2023, September 4). Facebook expansion: Invisible impacts? *The Shores of Academia*. www.shoresofacademia.substack.com/p/facebook-expansion-invisible-impacts

Steinberg, L. (2023). *Adolescence* (13th ed.). McGraw Hill.

Stevens, M. W. R., Dorstyn, D., Delfabbro, P. H., & King, D. L. (2021). Global prevalence of gaming disorder: A systematic review and meta-analysis. *Australian and New Zealand Journal of Psychiatry*, *55*(6), 553–568. doi.org/10.1177/0004867420962851

Su, R., Rounds, J., & Armstrong, P. I. (2009). Men and things, women and people: A meta-analysis of sex differences in interests. *Psychological Bulletin*, *135*(6), 859–884. doi.org/10.1037/a0017364

Su, W., Han, X., Yu, H., Wu, Y., & Potenza, M. N. (2020). Do men become addicted to internet gaming and women to social media? A meta-analysis examining gender-related differences in specific internet addiction. *Computers in Human Behavior*, *113*, 106480. doi.org/10.1016/j.chb.2020.106480

Substance Abuse and Mental Health Services Administration. (2023, January 4). *2021 NSDUH detailed tables*. www.samhsa.gov/data/report/2021-nsduh-detailed-tables

Sun, C., Bridges, A., Johnson, J. A., & Ezzell, M. B. (2016). Pornography and the male sexual script: An analysis of consumption and sexual relations. *Archives of Sexual Behavior*, *45*(4), 983–994. doi.org/10.1007/s10508-014-0391-2

Szuhany, K. L., & Simon, N. M. (2022). Anxiety disorders: A review. *JAMA*, *328*(24), 2431–2445. doi.org/10.1001/jama.2022.22744

Szymanski, D. M., & Stewart-Richardson, D. N. (2014). Psychological, relational, and sexual correlates of pornography use on young adult heterosexual men in romantic relationships. *Journal of Men's Studies*, *22*(1), 64–82. doi.org/10.3149/jms.2201.64

Taleb, N. N. (2012). *Antifragile: Things That Gain from Disorder*. Random House.

Tamana, S. K., Ezeugwu, V., Chikuma, J., Lefebvre, D. L., Azad, M. B., Moraes, T. J., Subbarao, P., Becker, A. B., Turvey, S. E., Sears, M. R., Dick, B. D., Carson, V., Rasmussen, C., CHILD Study Investigators, Pei, J., & Mandhane, P. J. (2019). Screen-time is associated with inattention problems in preschoolers: Results from the

CHILD birth cohort study. *PLoS ONE, 14*(4), Article e0213995. doi.org/10.1371/journal.pone.0213995

Tanil, C. T., & Yong, M. H. (2020). Mobile phones: The effect of its presence on learning and memory. *PLoS ONE, 15*(8), Article e0219233. doi.org/10.1371/journal.pone.0219233

Tannen, D. (1990). *You just don't understand: Women and men in conversation.* Ballantine Books.

Tanner, J. M. (1990). *Fetus into man: Physical growth from conception to maturity.* Harvard University Press.

Tarokh, L., Saletin, J. M., & Carskadon, M. A. (2016). Sleep in adolescence: Physiology, cognition, and mental health. *Neuroscience and Biobehavioral Reviews, 70,* 182–188. doi.org/10.1016/j.neubiorev.2016.08.008

Teo, A. R., & Gaw, A. C. (2010). Hikikomori, a Japanese culture-bound syndrome of social withdrawal? *Journal of Nervous and Mental Disease, 198*(6), 444–449. doi.org/10.1097/nmd.0b013e3181e086b1

Thompson, L., Sarovic, D., Wilson, P., Sämfjord, A., & Gillberg, C. (2022). A PRISMA systematic review of adolescent gender dysphoria literature: 1) Epidemiology. *PLoS Global Public Health, 2*(3), Article e0000245. doi.org/10.1371/journal.pgph.0000245

Thoreau, H. D. (1910). *Walden* (C. Johnsen, Illus.). Thomas Y. Crowell.

Thorn & Benenson Strategy Group. (2021, May). *Responding to online threats: Minors' perspectives on disclosing, reporting, and blocking.* info.thorn.org/hubfs/Research/Responding%20to%20Online%20Threats_2021-Full-Report.pdf

Thorndike, E. L. (1898). Animal intelligence: An experimental study of the associative processes in animals. *Psychological Review: Monograph Supplements, 2*(4), i–109. doi.org/10.1037/h0092987

Tierney, J., & Baumeister, R. F. (2019). *The power of bad: How the negativity effect rules us and how we can rule it.* Penguin Books.

Tomasello, M. (1994). The question of chimpanzee culture. In R. W. Wrangham, W. C. McGrew, F. B. M. de Waal, & P. G. Heltne (Eds.), *Chimpanzee cultures* (pp. 301–317). Harvard University Press.

Torre, M. (2018). Stopgappers? The occupational trajectories of men in female-dominated occupations. *Work and Occupations, 45*(3), 283–312. doi.org/10.1177/0730888418780433

Turban, J. L., Dolotina, B., King, D., & Keuroghlian, A. S. (2022). Sex assigned at birth ratio among transgender and gender diverse adolescents in the United States. *Pediatrics, 150*(3). doi.org/10.1542/peds.2022-056567

Turban, J. L., & Ehrensaft, D. (2018). Research review: Gender identity in youth: Treatment paradigms and controversies. *Journal of Child Psychology and Psychiatry, 59*(12), 1228–1243. doi.org/10.1111/jcpp.12833

Turkle, S. (2015). *Reclaiming conversation: The power of talk in a digital age.* Penguin.

Twenge, J. M. (2017). *iGen: Why today's super-connected kids are growing up less rebellious, more tolerant, less happy—and completely unprepared for adulthood—and what that means for the rest of us.* Atria Books.

Twenge, J. M. (2023, October 24). Here are 13 other explanations for the adolescent mental health crisis. None of them work. *After Babel.* www.afterbabel.com/p/13-explanations-mental-health-crisis

Twenge, J. M. (2023a). *Generations: The real differences between Gen Z, Millennials, Gen X, Boomers, and Silents—and what they mean for America's future.* Atria Books.

Twenge, J. M. (2023b). The mental health crisis has hit millennials. *After Babel.* www.afterbabel.com/p/the-mental-illness-crisis-millenials

Twenge, J. M., Gentile, B., DeWall, C. N., Ma, D., Lacefield, K., & Schurtz, D. R. (2010). Birth cohort increases in psychopathology among young Americans, 1938–2007: A cross-temporal meta-analysis of the MMPI. *Clinical Psychology Review, 30*(2), 145–154. doi.org/10.1016/j.cpr.2009.10.005

Twenge, J. M., Haidt, J., Blake, A. B., McAllister, C., Lemon, H. & Le Roy, A. (2021). Worldwide increases in adolescent loneliness. *Journal of Adolescence, 93*(1), 257–269. doi.org/10.1016/j.adolescence.2021.06.006

Twenge, J. M., Haidt, J., Lozano, J., & Cummins, K. M. (2022). Specification curve analysis shows that social media use is linked to poor mental health, especially among girls. *Acta Psychologica, 224*, 103512. doi.org/10.1016/j.actpsy.2022.103512

Twenge, J. M., Martin, G. N., & Campbell, W. K. (2018). Decreases in psychological well-being among American adolescents after 2012 and links to screen time during the rise of smartphone technology. *Emotion, 18*(6), 765–780. doi.org/10.1037/emo0000403

Twenge, J. M., Martin, G. N., & Spitzberg, B. H. (2019). Trends in U.S. adolescents' media use, 1976–2016: The rise of digital media, the decline of TV, and the (near) demise of print. *Psychology of Popular Media Culture, 8*(4), 329–345. doi.org/10.1037/ppm0000203

Twenge, J. M., Spitzberg, B. H., & Campbell, W. K. (2019). Less in-person social interaction with peers among U.S. adolescents in the 21st century and links to loneliness. *Journal of Social and Personal Relationships, 36*(6), 1892–1913. doi.org/10.1177/0265407519836170

Twenge, J., Wang, W., Erickson, J., & Wilcox, B. (2022). Teens and tech: What difference does family structure make? *Institute for Family Studies/Wheatley Institute.* www.ifstudies.org/ifs-admin/resources/reports/teensandtech-final-1.pdf

Twenge, J. M., Zhang, L., & Im, C. (2004). It's beyond my control: A cross-temporal meta-analysis of increasing externality in locus of control, 1960–2002. *Personality and Social Psychology Review, 8*(3), 308–319. doi.org/10.1207/s15327957pspr0803_5

Uhls, Y. T., Ellison, N. B., & Subrahmanyam, K. (2017). Benefits and costs of social media in adolescence. *Pediatrics, 140*(Supplement 2), S67–S70. doi.org/10.1542/peds.2016-1758e

U.S. Bureau of Labor Statistics. (n.d.). *Civilian unemployment rate.* www.bls.gov/charts/employment-situation/civilian-unemployment-rate.htm

U.S. Department of Health and Human Services. (2023). *Social media and youth mental health: The U.S. surgeon general's advisory.* www.hhs.gov/surgeongeneral/priorities/youth-mental-health/social-media/index.html

Vaillancourt-Morel, M.-P., Blais-Lecours, S., Labadie, C., Bergeron, S., Sabourin, S., & Godbout, N. (2017). Profiles of cyberpornography use and sexual well-being in adults. *Journal of Sexual Medicine, 14*(1), 78–85. doi.org/10.1016/j.jsxm.2016.10.016

van Elk, M., Arciniegas Gomez, M. A., van der Zwaag, W., van Schie, H. T., & Sauter, D. (2019). The neural correlates of the awe experience: Reduced default mode network activity during feelings of awe. *Human Brain Mapping, 40*(12), 3561–3574. doi.org/10.1002/hbm.24616

Vella-Brodrick, D. A., & Gilowska, K. (2022). Effects of nature (greenspace) on cognitive functioning in school children and adolescents: A systematic review. *Educational Psychology Review, 34*(3), 1217–1254. doi.org/10.1007/s10648-022 -09658-5

Verduyn, P., Lee, D. S., Park, J., Shablack, H., Orvell, A., Bayer, J., Ybarra, O., Jonides, J., & Kross, E. (2015). Passive Facebook usage undermines affective well-being: Experimental and longitudinal evidence. *Journal of Experimental Psychology: General, 144*(2), 480–488. doi.org/10.1037/xge0000057

Vermeulen, K. (2021). *Generation disaster: Coming of age post-9/11.* Oxford University Press.

Viner, R., Davie, M., & Firth, A. (2019). *The health impacts of screen time: A guide for clinicians and parents.* Royal College of Paediatrics and Child Health. www .rcpch.ac.uk/sites/default/files/2018-12/rcpch_screen_time_guide_-_final.pdf

Vogels, E. A. (2021, June 22). Digital divide persists even as Americans with lower incomes make gains in tech adoption. Pew Research Center. www.pewresearch .org/short-reads/2021/06/22/digital-divide-persists-even-as-americans-with -lower-incomes-make-gains-in-tech-adoption/

Vogels, E. A. (2022, December 15). Teens and cyberbullying 2022. Pew Research Center. www.pewresearch.org/internet/2022/12/15/teens-and-cyberbullying-2022/

Vogels, E. A., & Gelles-Watnick, R. (2023, April 24). Teens and social media: Key findings from Pew Research Center surveys. Pew Research Center. www.pewresearch .org/short-reads/2023/04/24/teens-and-social-media-key-findings-from-pew -research-center-surveys/

Vogels, E. A., Gelles-Watnick, R., & Massarat, N. (2022, August 10). Teens, social media, and technology 2022. Pew Research Center. www.pewresearch.org/internet/2022 /08/10/teens-social-media-and-technology-2022/

Wagner, S., Panagiotakopoulos, L., Nash, R., Bradlyn, A., Getahun, D., Lash, T. L., Roblin, D., Silverberg, M. J., Tangpricha, V., Vupputuri, S., & Goodman, M. (2021). Progression of gender dysphoria in children and adolescents: A longitudinal study. *Pediatrics, 148*(1), Article e2020027722. doi.org/10.1542/peds.2020-027722

Walker, R. J., Hill, K., Burger, O. F., & Hurtado, A. (2006). Life in the slow lane revisited: Ontogenetic separation between chimpanzees and humans. *American Journal of Physical Anthropology, 129*(4), 577–583. doi.org/10.1002/ajpa.20306

Waller, J. (2008). *A time to dance, a time to die: The extraordinary story of the dancing plague of 1518.* Icon Books.

Wang, L., Zhou, X., Song, X., Gan, X., Zhang, R., Liu, X., Xu, T., Jiao, G., Ferraro, S., Bore, M. C., Yu, F., Zhao, W., Montag, C., & Becker, B. (2023). Fear of missing out (FOMO) associates with reduced cortical thickness in core regions of the posterior default mode network and higher levels of problematic smartphone and social media use. *Addictive Behaviors, 143*, 107709. doi.org/10.1016/j.addbeh.2023 .107709

Ward, A. F., Duke, K., Gneezy, A., & Bos, M. W. (2017). Brain drain: The mere presence of one's own smartphone reduces available cognitive capacity. *Journal of the Association for Consumer Research, 2*(2), 140–154. doi.org/10.1086/691462

Wass, S. V., Whitehorn, M., Marriott Haresign, I., Phillips, E., & Leong, V. (2020). Interpersonal neural entrainment during early social interaction. *Trends in Cognitive Sciences, 24*(4), 329–342. doi.org/10.1016/j.tics.2020.01.006

Webb, C. (2016). *How to have a good day: Harness the power of behavioral science to transform your working life.* National Geographic Books.

Wessely, S. (1987). Mass hysteria: Two syndromes? *Psychological Medicine, 17*(1), 109–120. doi.org/10.1017/S0033291700013027

Wheaton, A. G., Olsen, E. O., Miller, G. F., & Croft, J. B. (2016). Sleep duration and injury-related risk behaviors among high school students—United States, 2007–2013. *Morbidity and Mortality Weekly Report, 65*(13), 337–341. www.jstor.org/stable /24858002

Wiedemann, K. (2015). Anxiety and anxiety disorders. In *International Encyclopedia of the Social and Behavioral Sciences*, 804–810. doi.org/10.1016/B978-0-08-097086 -8.27006-2

Willoughby, B. J., Carroll, J. S., Busby, D. M., & Brown, C. C. (2016). Differences in pornography use among couples: Associations with satisfaction, stability, and relationship processes. *Archives of Sexual Behavior, 45*(1), 145-158. doi.org/10.1007 /s10508-015-0562-9

Wilson, D. S. (2002). *Darwin's cathedral: Evolution, religion, and the nature of society.* University of Chicago Press.

Wilson, E. O. (1984). *Biophilia: The human bond with other species.* Harvard University Press.

Wilson, S. J., & Lipsey, M. W. (2000). Wilderness challenge programs for delinquent youth: A meta-analysis of outcome evaluations. *Evaluation and Program Planning, 23*(1), 1–12. doi.org/10.1016/S0149-7189(99)00040-3

Wiltermuth, S. S., & Heath, C. (2009). Synchrony and cooperation. *Psychological Science, 20*(1), 1–5. doi.org/10.1111/j.1467-9280.2008.02253.x

Wittek, C. T., Finserås, T. R., Pallesen, S., Mentzoni, R. A., Hanss, D., Griffiths, M. D., & Molde, H. (2016). Prevalence and predictors of video game addiction: A study based on a national representative sample of gamers. *International Journal of Mental Health and Addiction, 14*, 672–686. doi.org/10.1007/s11469-015-9592-8

Wolfson, A. R., & Carskadon, M. A. (2003). Understanding adolescents' sleep patterns and school performance: A critical appraisal. *Sleep Medicine Reviews, 7*(6), 491–506. doi.org/10.1016/s1087-0792(03)90003-7

Wright, P. J., Tokunaga, R. S., Kraus, A., & Klann, E. (2017). Pornography consumption and satisfaction: A meta-analysis. *Human Communication Research, 43*(3), 315–343. doi.org/10.1111/hcre.12108

Young, D. R., McKenzie, T. L., Eng, S., Talarowski, M., Han, B., Williamson, S., Galfond, E., & Cohen, D. A. (2023). Playground location and patterns of use. *Journal of Urban Health, 100*(3), 504–512. doi.org/10.1007/s11524-023-00729-8

Young, K. (2009). Understanding online gaming addiction and treatment issues for adolescents. *American Journal of Family Therapy, 37*(5), 355–372. doi.org/10.1080 /01926180902942191

Zahn-Waxler, C., Shirtcliff, E. A., & Marceau, K. (2008). Disorders of childhood and adolescence: Gender and psychopathology. *Annual Review of Clinical Psychology, 4*(1), 275–303. doi.org/10.1146/annurev.clinpsy.3.022806.091358

Zastrow, M. (2017). Is video game addiction really an addiction? *Proceedings of the National Academy of Sciences, 114*(17), 4268–4272. doi.org/10.1073/pnas.1705077114

Zeanah, C. H., Gunnar, M. R., McCall, R. B., Kreppner, J. M., & Fox, N. A. (2011). Sensitive periods. *Monographs of the Society for Research in Child Development, 76*(4), 147–162. doi.org/10.1111/j.1540-5834.2011.00631.x

Zendle, D., & Cairns, P. (2018). Video game loot boxes are linked to problem gambling: Results of a large-scale survey. *PLoS ONE, 13*(11), Article e0206767. doi.org/10.1371/journal.pone.0206767

Zucker, K. J. (2019). Adolescents with gender dysphoria: Reflections on some contemporary clinical and research issues. *Archives of Sexual Behavior, 48,* 1983–1992. DOI: 10.1007/s10508-019-01518-8

Zucker, K. J. (2017). Epidemiology of gender dysphoria and transgender identity. *Sexual Health, 14*(5), 404–411. doi.org/10.1071/sh17067

INDEX

Page numbers in *italics* refer to figures.